AP* PHYSICS C
ELECTRICITY AND MAGNETISM

2020 Edition

100 MUST-KNOW QUESTIONS IN
1. ELECTROSTATICS
2. CONDUCTORS, CAPACITORS, DIELECTRICS

With Answers and Explanations

Sudhir K. Sood, Ph.D.

* AP, Advanced Placement Program, and College Board are registered trademarks of the College Board, which was not involved in the production of, and does not endorse, this book.

© Copyright 2020 by Sudhir K. Sood, Ph.D.
All rights reserved. No part of this book may be reproduced, stored in retrieval system or transmitted in any form or by any means whatsoever except as permitted by applicable copyright law.

Preface

'*AP Physics C Electricity and Magnetism, 2020 Edition: 100 Must-Know Questions in 1. Electrostatics 2. Conductors, Capacitors, Dielectrics With Answers and Explanations*' has been written primarily to help students master the concepts and techniques they need to earn highest possible score in AP Physics C: Electricity and Magnetism exam.

This book contains 90 Multiple Choice and 10 Free Response Questions pertaining to the aforementioned two units that together comprise 40-51% of the complete AP Physics C: Electricity and Magnetism exam. Using this book, students are sure to gain mastery over every type of question that they are ever likely to find in the exam. This becomes plausible because of the judicious way the book is laid out. First step that assures complete coverage is the division of the two units into topics (eight in all) that coincide with those specified in the Course Framework updated by the College Board for the 2019-20 school year. Secondly, careful analysis of the exam questions and related information issued by the College Board from time to time coupled with vast teaching experience of the author has assured the inclusion of virtually all question-types for each of these eight topics. The division of questions into topics has the added advantage of allowing the students to easily find and improve upon those parts that they find difficult to grasp.

Additionally, the book also provides Answer Key and Detailed Explanations for each of the questions. For answering the Physics C exam questions, students require a far deeper understanding of the concepts as compared to other easier exams, where, quite often, knowledge of the final results alone suffices. Keeping this in mind, we have always included, at appropriate places, complete derivations of the result being used to arrive at the answer. This will also help students recall an important component of the theory part that they would have studied otherwise.

The book can be used by the students either as a self-study guidebook or as a valuable supplementary text to AP classroom courses. It may also be used as perfect review material just before the exam.

Even though the book is designed for AP Physics C exam, it can be equally useful for students taking calculus-based Physics courses.

Any communication from students and teachers regarding deficiencies of the book and suggestions for improvement are welcome. Your help in pointing out errors that might have escaped attention despite effort to produce error free manuscript will be greatly appreciated. You can contact me at sudhbapc@ gmail.com

AP Physics C: Electricity and Magnetism Course Updates for the 2019-20 School Year - Issued by the College Board
(Part included in this Book)

UNIT 1: Electrostatics

Exam Weighting for the Multiple-Choice Section of the AP Exam: 26–34%

   Topics

      1.1 Charge and Coulomb's Law
      1.2 Electric Field and Electric Potential
      1.3 Potential Due to Point Charges and Uniform Fields
      1.4 Gauss's Law
      1.5 Fields and Potentials of Other Charge Distributions

UNIT 2: Conductors, Capacitors, Dielectrics

Exam Weighting for the Multiple-Choice Section of the AP Exam: 14–17%

      2.1 Electrostatics with Conductors
      2.2 Capacitors
      2.3 Dielectrics

About the Author

Sudhir K. Sood earned his Ph.D. degree in fundamental particle physics from University of Delhi. Subsequently, as research scientist and Professor of Physics at Universities in France, Canada and India, Dr. Sood has taught a number of courses both at introductory and advanced graduate level. He has lectured at international Physics conferences and authored numerous well-cited research papers that are published in reputed peer reviewed journals. More recently, for more than a decade, he has taught students in Delhi who wish to specialize in engineering, medicine and physical science courses.

Other Titles by the Author

1. High School Physics: Master It With Ease (1) Introductory Current Electricity
2. High School Physics: Master It With Ease (2) Introductory Electromagnetism
3. High School Physics: Master It With Ease (3) Introduction To Reflection and Refraction of Light, Mirrors and Lenses
4. A to …Z Class 10 CBSE/NCERT Physics (For Indian students)

Contents

PREFACE 3

CONTENTS 5

PART I: QUESTIONS 6

   A. MULTIPLE-CHOICE QUESTIONS 7

      UNIT 1: Electrostatics 7

         1.1 Charge and Coulomb's Law 7
         1.2 Electric Field and Electric Potential *and* 11
         1.3 Electric Potential Due to Point Charges and Uniform Fields 11
         1.4 Gauss's law 17
         1.5 Fields and potentials of other charge distributions 23

      UNIT 2: Conductors, Capacitors, Dielectrics 25

         2.1 Electrostatics with Conductors 25
         2.2 Capacitors 31
         2.3 Dielectrics 34

   B. FREE-RESPONSE QUESTIONS 37-41

PART II: ANSWERS 42

ANSWER KEY 43

ANSWERS AND EXPLANATIONS 44

   A. MULTIPLE-CHOICE QUESTIONS 44

      UNIT 1: Electrostatics 44

         1.1 Charge and Coulomb's Law 44
         1.2 Electric Field and Electric Potential *and* 52
         1.3 Electric Potential Due to Point Charges and Uniform Fields 52
         1.4 Gauss's law 60
         1.5 Fields and potentials of other charge distributions 72

      UNIT 2: Conductors, Capacitors, Dielectrics 78

         2.1 Electrostatics with Conductors 78
         2.2 Capacitors 86
         2.3 Dielectrics 96

   B. FREE-RESPONSE QUESTIONS 100-126

PART I: QUESTIONS

[In what follows, $k = 1/(4\pi\varepsilon_0) = 9 \times 10^9$ N m² C⁻²]

A. MULTIPLE-CHOICE QUESTIONS

UNIT 1: Electrostatics

1.1 Charge and Coulomb's Law

1. Suppose F is the magnitude of force of repulsion between two positively charged particles, each having charge q and separated by distance r. If e denotes the charge of an electron, the number n of electrons missing from each particle is directly proportional to

   (A) F
   (B) F^2
   (C) $\sqrt{F}$
   (D) $1/F$
   (E) $1/\sqrt{F}$

2. A positive charge q_1 exerts some electrostatic force $\mathbf{F}$ on a second positive charge q_2. A third charge q_3 is brought close to q_1. Which of the following statement is true?

   (A) Magnitude of $\mathbf{F}$ will decrease but its direction will not change.
   (B) $\mathbf{F}$ will remain unchanged.
   (C) Magnitude of $\mathbf{F}$ will increase but its direction will not change.
   (D) Magnitude of $\mathbf{F}$ will increase if q_3 is of the same sign as q_1 and will decrease if q_3 is of opposite sign but direction of $\mathbf{F}$ will not change.
   (E) Direction of $\mathbf{F}$ will change but its magnitude will not change.

3. Two particles having equal charge q and equal mass m are placed separated by some distance. If the electrostatic force between them is balanced by the gravitational force, the ratio (q/m) is

   (A) $\dfrac{1}{\sqrt{4\pi\varepsilon_0 G}}$
   (B) $\sqrt{4\pi\varepsilon_0 G}$
   (C) $4\pi\varepsilon_0 G$
   (D) $\dfrac{G}{4\pi\varepsilon_0}$
   (E) $\dfrac{4\pi\varepsilon_0}{G}$

4. Two equal point charges q are placed at $x = -a$ and $x = +a$ on the x-axis. Another point charge Q is placed at the origin. If the system of three charges is in equilibrium, the value of Q is

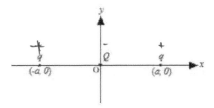

   (A) $Q = -2q$
   (B) $Q = q/4$
   (C) $Q = -q/2$
   (D) $Q = -q/4$
   (E) $Q = -4q$

Questions 5-7

Three charges $-q, Q$ and $-2q$ are placed at the origin $O\,(0,0)$ and the points $A\,(a, 0)$ and $B\,(-a, 2a)$ respectively as shown.

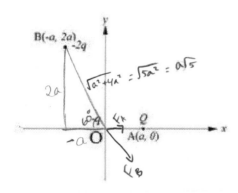

5. The x-component of the net force on $-q$ is proportional to

   (A) $\dfrac{Q}{a^2} - \dfrac{q}{5\sqrt{5}\,a^2}$

   (B) $\dfrac{q}{5\sqrt{5}\,a^2}$

   (C) $\dfrac{Q}{a^2} + \dfrac{q}{5\sqrt{5}\,a^2}$

   (D) $\dfrac{Q}{a^2} - \dfrac{2q}{5\sqrt{5}\,a^2}$

   (E) $\dfrac{Q}{a^2} + \dfrac{2q}{5\sqrt{5}\,a^2}$

6. The y-component of the net force on $-q$ is

   (A) proportional to Q
   (B) independent of q
   (C) independent of Q
   (D) proportional to Q^2
   (E) None of the above

7. The net force **F** on $-q$ is

   (A) $\dfrac{q}{4\pi\varepsilon_0}\left(\dfrac{Q}{a^2} + \dfrac{q}{5\sqrt{5}\,a^2}\right)\mathbf{i} + \dfrac{1}{4\pi\varepsilon_0}\dfrac{2q^2}{5\sqrt{5}\,a^2}\mathbf{j}$

   (B) $\dfrac{q}{4\pi\varepsilon_0}\left(\dfrac{Q}{a^2} + \dfrac{2q}{5\sqrt{5}\,a^2}\right)\mathbf{i} - \dfrac{q^2}{\pi\varepsilon_0}\dfrac{1}{5\sqrt{5}\,a^2}\mathbf{j}$

   (C) $\dfrac{q}{4\pi\varepsilon_0}\left(\dfrac{Q}{a^2} + \dfrac{q\,a}{(5\sqrt{5}\,a^2)^{\tfrac{3}{2}}}\right)\mathbf{i} - \dfrac{1}{4\pi\varepsilon_0}\dfrac{2a\,q}{(\sqrt{5}\,a)^3}\mathbf{j}$

   (D) $\dfrac{q}{4\pi\varepsilon_0}\left(\dfrac{Q}{a^2} + \dfrac{q}{5\sqrt{5}\,a^2}\right)\mathbf{i} - \dfrac{1}{4\pi\varepsilon_0}\dfrac{2q^2}{5\sqrt{5}\,a^2}\mathbf{j}$

   (E) $\dfrac{q}{4\pi\varepsilon_0}\left(\dfrac{Q}{a^2} + \dfrac{q}{5\sqrt{5}\,a^2}\right)\mathbf{i} + \dfrac{1}{4\pi\varepsilon_0}\dfrac{2q}{5\sqrt{5}\,a^2}\mathbf{j}$

8. Charge Q is placed at each of the three corners A, B and C of a square ABCD of side a. A charge q is placed at the fourth corner of the square (see diagram given below). The ratio of the magnitude of resultant electrical force on the charge q due to three charges at A, B and C to the magnitude of electrical force on the charge q due to single charge at A is

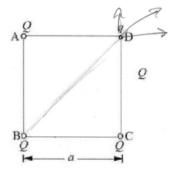

   (A) 2
   (B) $\sqrt{2} + 1/2$
   (C) $\dfrac{1}{\sqrt{2}}$
   (D) $2 + \dfrac{1}{\sqrt{2}}$
   (E) $1/2$

9. Two equally charged identical metal spheres A and B of negligible size, when placed at a distance r from each other, repel each other with an electrostatic force of magnitude 10^{-5} N. Another identical uncharged sphere C is touched to A and then placed at the mid-point between A and B. What is the net force on C?

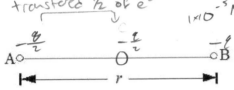

(A) $(1/4) \times 10^{-5}$ N along CB.

(B) 4×10^{-5} N along CB.

(C) 2×10^{-5} N along CA.

(D) 10^{-5} N along CB.

(E) 10^{-5} N along CA.

10. A charge Q is divided into two parts. These parts are then placed at some separation d. The force between them will be maximum if the values of two parts are

(A) $\left(\dfrac{Q}{4}, \dfrac{3Q}{4}\right)$

(B) $\left(\dfrac{Q}{3}, \dfrac{2Q}{3}\right)$

(C) $((0, Q))$

(D) $\left(\dfrac{Q}{2}, \dfrac{Q}{2}\right)$

(E) None of the above

11. A charge Q is fixed at the origin O $(0, 0)$ and another charge q is placed at a point A $(x_0, 0)$ on a smooth plane surface. It starts moving under the electrostatic force of repulsion between the two charges. If v is its velocity when it is at a point B $(x_1, 0)$, we have

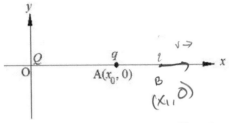

(A) $\dfrac{1}{v} = \dfrac{1}{4\pi\varepsilon_0} \dfrac{qQ}{m} \left[\dfrac{1}{x_0} - \dfrac{1}{x_1}\right]$

(B) $v = \dfrac{1}{4\pi\varepsilon_0} \dfrac{qQ}{m} \left[\dfrac{1}{x_0} - \dfrac{1}{x_1}\right]$

(C) $v = \dfrac{1}{4\pi\varepsilon_0} \dfrac{qQ}{m} \left[\dfrac{1}{x_0}\right]$

(D) $v^2 = \dfrac{1}{2\pi\varepsilon_0} \dfrac{qQ}{m} \left[\dfrac{1}{x_0} - \dfrac{1}{x_1}\right]$

(E) $v^2 = \dfrac{1}{4\pi\varepsilon_0} \dfrac{qQ}{m} \left[\dfrac{1}{x_1} - \dfrac{1}{x_0}\right]$

12. As shown below, five identical charges of magnitude q are placed at each of the five corners A, B, C, D and E of a regular hexagon ABCDEF. If d is the distance of the center of the hexagon to each corner, the electric force on point charge $-q$ placed at the centre O of the hexagon is

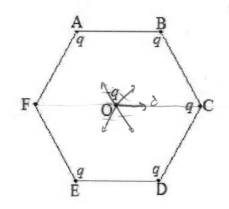

(A) $\dfrac{q^2}{20\pi\varepsilon_0 d^2}$ along OF.

(B) $\dfrac{q^2}{4\pi\varepsilon_0 d^2}$ along FO.

(C) $\dfrac{q^2}{20\pi\varepsilon_0 d^2}$ along FO.

(D) $\dfrac{q^2}{4\pi\varepsilon_0 d^2}$ along OF.

(E) $\dfrac{q^2}{8\pi\varepsilon_0 d^2}$ towards OF.

Questions 13-14

Consider a system of two charges, each equal to $+Q$, are kept at points A (0, a) and B (0, -a) respectively of an x-y coordinate system.

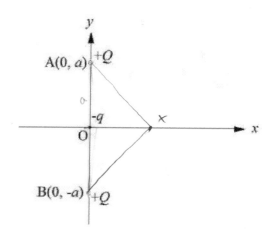

13. A particle having charge $-q$ is placed at the origin. If charge $-q$ is given a small displacement x ($\ll a$) along the x-axis, the magnitude of net force acting on the particle is proportional to
(A) x^2
(B) $x^{1/2}$
(C) x
(D) $1/x$
(E) $x^{3/2}$

14. A small particle with negative charge $-q$ is released from a point on the negative x-axis at large distance from the origin O. It moves along the x-axis, passes through O and moves far away from O along positive x-axis. The magnitude F of the electrostatic force on the negatively charged particle is plotted against its x-coordinate. Taking the positive direction of force to be along positive x-axis, the resulting plot showing F as a function of x is best represented by

$F = \dfrac{2kqQ}{(a^2+x^2)^{3/2}} x$

$F_e = qE$

$E = \dfrac{F_e}{q} = \dfrac{kq}{r^2} \hat{r}$

11

(A)

@ x=0, F=0

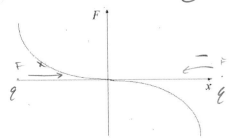

1.2 Electric Field and Electric Potential *and*

1.3 Electric Potential Due to Point Charges and Uniform Fields

15. A uniform electric field of magnitude 10^5 N/C exists in certain region. A positively charged oil drop of mass 5 µg is suspended in the field such that it neither falls nor rises. The charge Q on the drop will be (take $g = 10$ m s^{-2})
    (A) 5×10^{-13} C
    (B) 5×10^{-12} C
    (C) 5×10^{-14} C
    (D) 10^{-13} C
    (E) 5×10^{-10} C

$\Sigma F_e = ma$
$qE = mg$
$q \cdot 10^5 = 5 \times 10^{-9}(10)$
$q = 5 \times 10^{-13}$ C

(B)

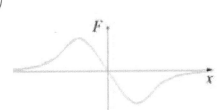

(C)

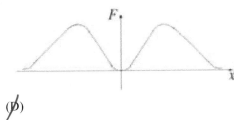

(D)

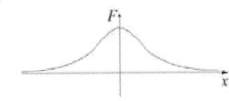

16. Three charges q, Q and $4q$ are placed on the x-axis at the points O (0, 0), A (a, 0) and B (2a, 0) respectively. The electric field at the origin O is zero if

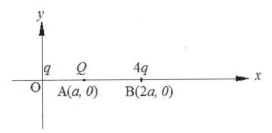

(A) $Q = -4q$
(B) $Q = -2q$
(C) $Q = -q$
(D) $q = -2Q$
(E) $q = -4Q$

$\dfrac{kQ}{a^2} = \dfrac{k(4q)}{4a^2}$

$Q = -q$

(E)

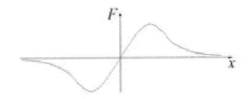

17. The figure given below shows three electric field lines. If F_P, F_Q and F_R are the magnitudes of forces on a test unit charge at the positions P, Q and R respectively, then

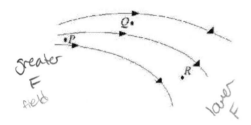

(A) $F_P < F_Q < F_R$
(B) $F_P > F_Q = F_R$
(C) $F_P = F_Q > F_R$
(D) $F_P < F_Q > F_R$
(E) $F_P > F_Q > F_R$

Questions 18-19

An electric dipole consisting of a pair of equal and opposite charges q and –q separated by a distance d is placed in a uniform external field E as shown in the figure given below. The line joining the charges makes an angle θ with direction of E.

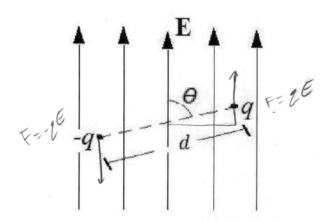

18. The net force exerted by the electric field on the dipole is

(A) $q\mathbf{E}$
(B) $2 q\mathbf{E}$
(C) zero
(D) $q\mathbf{E}/2$
(E) None of the above

19. The magnitude of torque on the dipole is

(A) $2q\, d\, E \sin \theta$
(B) $q\, d\, E \sin \theta$
(C) $q\, d\, E \cos \theta$
(D) is zero
(E) $q\, d\, E \tan \theta$

Questions 20-21

20. The pattern of electric field lines due to a system of two charges q_1 and q_2 fixed at two different points on the x-axis suggests that

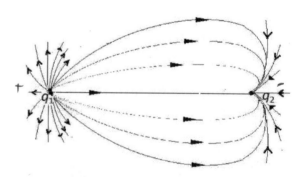

(A) q_1 is positive and q_2 is negative and $|q_1| > |q_2|$.
(B) Both q_1 and q_2 are negative and $|q_1| < |q_2|$.
(C) Both q_1 and q_2 are positive and $|q_1| < |q_2|$.
(D) q_1 is positive and q_2 is negative and $|q_1| = |q_2|$.
(E) q_1 is positive and q_2 is negative and $|q_1| < |q_2|$.

21. Electric field can be zero
    (A) at a finite distance to the left of q_1
    (B) at a finite distance to the right of q_2
    (C) at a finite distance to the right of q_1
    (D) at a finite distance to the left of q_2
    (E) at an infinite distance from both charges

22. A uniform electric field **E** along x-axis exists in certain region. If the work done in moving a charge of 0.4 C through a distance of 3 m along a line AB making an angle 60° with x-axis is 6 J, then the magnitude of electric field is

    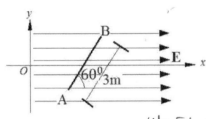

    (A) $\left(\frac{1}{\sqrt{3}}\right) \times 10.0$ N/C
    (B) 10.0 J/C
    (C) 10.0 N/C
    (D) 20.0 N/C
    (E) 5.0 N/C

23. The dimensions of potential difference are
    (A) $L^2 T^{-3} A^{-1}$
    (B) $M L^2 T^{-3} A^{-2}$
    (C) $M L^2 T^{-3} A^{-1}$
    (D) $M L^2 T^{-2} A^{-1}$
    (E) $M^2 L^{-2} T^{-3} A^{-1}$

24. Consider two point charges $-q$ and $+q$ that are placed at points A $(-a, 0)$ and B $(a, 0)$ respectively on x-axis. At the origin O (0, 0),

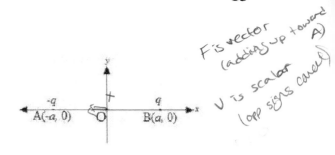

    (A) both electric field and electric potential are zero
    (B) electric field is zero but electric potential is not zero
    (C) electric field is not zero but electric potential is zero
    (D) neither electric field nor electric potential is zero
    (E) electric field is infinite but electric potential is not zero

25. Two charges q and $-q$ are placed at A $(-a, 0)$ and B $(a, 0)$ on the x-axis. The work done by an external force in moving a point charge Q from the origin O (0, 0) to C $(2a, 0)$ along the semicircle OSC is

    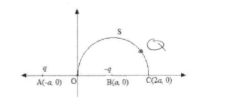

    (A) $\frac{Q q}{6 \pi \varepsilon_0 a}$
    (B) $-\frac{Q q}{6 \pi \varepsilon_0 a}$
    (C) $-\frac{Q q}{3 \pi \varepsilon_0 a}$
    (D) $-\frac{2 Q q}{3 \pi \varepsilon_0 a}$
    (E) $\frac{Q q}{3 \pi \varepsilon_0 a}$

26. A charged particle having charge Q is fixed at a point A on a frictionless horizontal surface. Another particle having charge q and mass m is shot along the surface from far away towards the first particle with speed v. The distance of closest approach d is

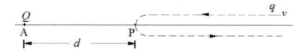

(A) $\dfrac{qQ}{8\pi\varepsilon_0 m v^2}$

(B) $\dfrac{Q}{2\pi\varepsilon_0 m v^2}$

(C) $\dfrac{2q}{4\pi\varepsilon_0 m v^2}$

(D) $\dfrac{qQ}{2\pi\varepsilon_0 m v^2}$

(E) $\dfrac{qQ}{4\pi\varepsilon_0 m v^2}$

27. Three point charges Q, $+q$ and $+q$ are placed, one at each corner of a right angled isosceles triangle ABC as shown in the figure. If the net electrostatic energy of the arrangement is zero, the ratio Q/q is equal to

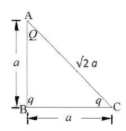

(A) $\dfrac{\sqrt{2}}{\sqrt{2}+1}$

(B) $-\dfrac{2}{\sqrt{2}+1}$

(C) $-\dfrac{\sqrt{2}}{\sqrt{2}+1}$

(D) $-\dfrac{\sqrt{2}}{\sqrt{2}+2}$

(E) $\dfrac{\sqrt{2}}{\sqrt{2}+2}$

28. Four charges q, $-q$, q and $-q$ are arranged at the corners PQRS of a square of side a, as shown below. The work done by external force in moving a point charge $+q_0$ from infinity to the centre O of the square is

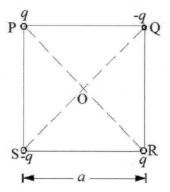

(A) 0

(B) $\dfrac{1}{2\pi\varepsilon_0}\dfrac{(q\,q_0)}{a}$

(C) $-\dfrac{1}{2\pi\varepsilon_0}\dfrac{(q\,q_0)}{a}$

(D) $\dfrac{1}{\pi\varepsilon_0}\dfrac{(q\,q_0)}{a}$

(E) $\dfrac{1}{4\pi\varepsilon_0}\dfrac{(q\,q_0)}{a}$

29. The variation of electric field E along the x-axis produced in certain region is given by the following graph.

Which of the following graphs best represents the variation of electric potential V along the x-axis?

(A)

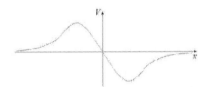

(B)

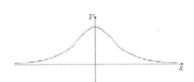

(C)

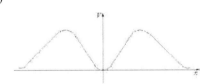

(D)

(E)

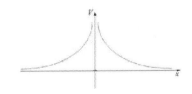

30. Some equipotential surfaces are shown below with the values of the electric potential written alongside. Among the points P, Q and R, the magnitude of the electric field is greatest at the point (s)

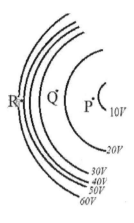

(A) P
(B) Q
(C) R
(D) Both P and
(E) None of the above

31. P, Q and R are three points in a uniform electric field **E** existing in a region. If V_P, V_Q and V_R are the values of electric potential at P, Q and R,

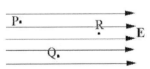

(A) $V_P < V_Q > V_R$
(B) $V_P < V_Q < V_R$
(C) $V_P = V_Q = V_R$
(D) $V_P > V_Q > V_R$
(E) $V_P > V_Q < V_R$

32. A uniform electric field **E** exists in certain region. A charged particle A having mass m and charge q, initially at rest, moves through a certain distance d in the field in time t_1. Another particle B having the same charge q but mass M, also initially at rest, moves through the same distance d in time t_2. The ratio t_1/t_2 is

(A) mM

(B) $\left(\dfrac{m}{M}\right)^{\frac{1}{2}}$

(C) $\left(\dfrac{M}{m}\right)^{\frac{1}{2}}$

(D) $\dfrac{m}{M}$

(E) $\dfrac{M}{m}$

33. A point charge q is moved from A $(a, 2a, 0)$ to the origin O $(0, 0, 0)$ along the path ABCO in a uniform electric field **E** along x-axis as shown in the figure given below. The coordinates of points B and C are $(2a, 0, 0)$ and $(a, -2a, 0)$ respectively. The work done by the electric field W_E is

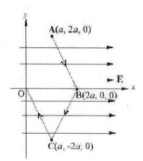

(A) $-2aqE$

(B) aqE

(C) $-aqE$

(D) $2aqE$

(E) zero

34. A particle of mass m and charge q is held at point A, which is at a distance of d from a fixed charge Q at point O. If this particle is released from rest, it moves away due to repulsion. Using energy concepts, find the speed v of particle when it is at point B, which is at a distance of $2d$ from the fixed charge.

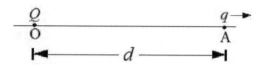

(A) $\left(\dfrac{qQ}{8\pi\varepsilon_0 m d}\right)^{\frac{1}{2}}$

(B) $\left(\dfrac{qQ}{4\pi\varepsilon_0 m d}\right)^{\frac{1}{2}}$

(C) $\dfrac{qQ}{4\pi\varepsilon_0 m d}$

(D) $\dfrac{qQ}{8\pi\varepsilon_0 m d}$

(E) $\left(\dfrac{qQ}{4\pi\varepsilon_0 m d}\right)^{2}$

35. In a region, the potential at any point (x, y, z) is represented by $V = 3x^2$, where V is in volts and x, y, z are in meters. The electric force experienced by a charge of 2.0 nC situated at (2m, 0, 1m) is

(A) 24.0×10^{-8} N **i**

(B) -24.0×10^{-9} N **i**

(C) -12.0×10^{-9} N **i**

(D) -2.0×10^{-9} N **i**

(E) 24.0×10^{-9} N **j**

36. As shown below, a circle of radius R is drawn with charge q at the centre O. Which of the following statement is correct?

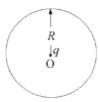

(A) the circle is an equipotential surface
(B) the circle is not an equipotential surface
(C) work done in carrying a charge q_0 from point B to C along the circle is not zero
(D) electric field at any point on the circle is zero
(E) electric field at any point on the circle is along tangent to the circle

1.4 Gauss's law $\phi = \oint \vec{E} \cdot d\vec{A} = \frac{q_{in}}{\varepsilon_0}$

37. Consider a square frame of side 20 cm placed in a uniform electric field $\mathbf{E} = 40.0\,\mathbf{i}$ V/m. If the plane of the square makes an angle of 30° angle with the x-axis, the flux φ through the square is

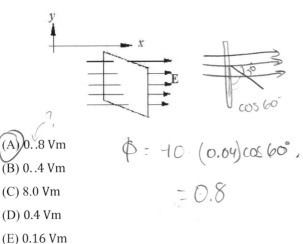

(A) 0..8 Vm
(B) 0..4 Vm
(C) 8.0 Vm
(D) 0.4 Vm
(E) 0.16 Vm

38. The electric field in a region is given as $\mathbf{E} = (E_0\,x)\,\mathbf{i}$ where E_0 is a constant. Consider a cubical volume of edge a bounded by the surfaces $x = 0$, $x = a$, $y = 0$, $y = a$, $z = 0$ and $z = a$ as shown in the diagram. The charge q contained inside this cubical volume is

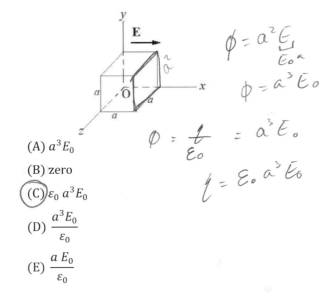

(A) $a^3 E_0$
(B) zero
(C) $\varepsilon_0\, a^3 E_0$
(D) $\dfrac{a^3 E_0}{\varepsilon_0}$
(E) $\dfrac{a\, E_0}{\varepsilon_0}$

39. A charge q is placed at the center of the mouth of the open end of a cylindrical vessel as shown below. The electric flux φ_c through the vessel is

(A) $\varepsilon_0 q$
(B) $\dfrac{q}{2\varepsilon_0}$
(C) $\dfrac{q}{\varepsilon_0}$
(D) $\dfrac{2q}{\varepsilon_0}$
(E) $\dfrac{\varepsilon_0}{2q}$

40. A charge q is situated at a distance $a/2$ above a square of side a. What is the electric flux φ_s through the square?

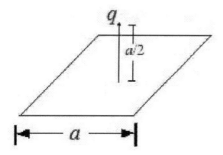

(A) $\dfrac{\varepsilon_0}{6\,q}$

(B) $6\,\varepsilon_0 q$

(C) $\dfrac{q}{6\,\varepsilon_0}$

(D) $\dfrac{q}{3\,\varepsilon_0}$

(E) $\dfrac{6\,q}{\varepsilon_0}$

41. A spherical Gaussian surface (shown by dotted line) and three point charges are shown in the figure given below. While applying Gauss's law, the electric field on the surface will be

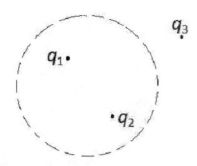

(A) due to q_1 and q_2 only
(B) due to q_3 only
(C) due to q_1, q_2 and q_3

(D) due to positive charges among the three
(E) none of the above

42. If the electric flux entering and leaving a closed surface is φ_1 and φ_2 respectively, the electric charge inside the surface is

(A) $\varepsilon_0(\varphi_1 - \varphi_2)$
(B) $\varepsilon_0(\varphi_2 - \varphi_1)$
(C) $\varepsilon_0 \varphi_2$
(D) $\dfrac{(\varphi_2 - \varphi_1)}{\varepsilon_0}$
(E) $\dfrac{(\varphi_2 - \varphi_1)}{2\varepsilon_0}$

43. The figure given below shows four closed surfaces S_1, S_2, S_3 and S_4 respectively together with the corresponding charge distributions. If $\varphi_1, \varphi_2, \varphi_3$ and φ_4 be respective electric fluxes through the surfaces

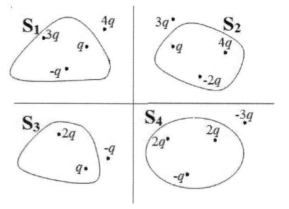

(A) $\varphi_1 < \varphi_2 < \varphi_3 < \varphi_4$
(B) $\varphi_1 = \varphi_2 = \varphi_3 = \varphi_4$
(C) $\varphi_1 > \varphi_2 > \varphi_3 > \varphi_4$
(D) $\varphi_1 > \varphi_2 = \varphi_3 < \varphi_4$
(E) $\varphi_1 > \varphi_2 > \varphi_3 = \varphi_4$

44. A charge q is enclosed by a spherical Gaussian surface of radius R. If the radius is doubled, then the outward electric flux will

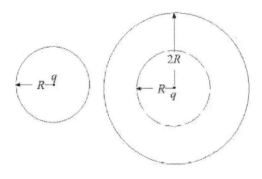

(A) remain unchanged
(B) increase four times
(C) decrease four times
(D) be reduced to half its earlier value
(E) be doubled

Questions 45-46

Consider a charged spherical shell of radius R having a total positive charge Q spread uniformly over its outer surface.

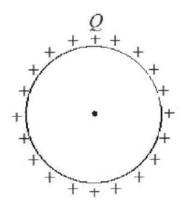

45. The variation of magnitude of electric field with the distance r from the centre is best described by

(A)
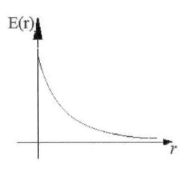

(B)
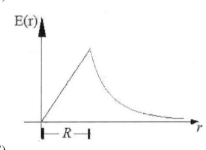

(C)
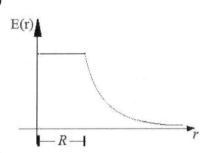

(D)
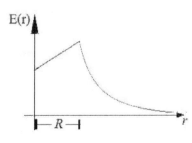

(E)
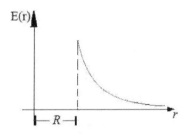

46. The variation of the electric potential $V(r)$ the distance r from the centre is best represented by which graph?

(A)
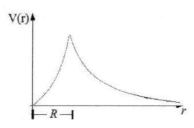

(B)
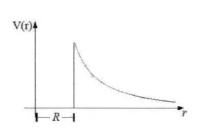

(C)
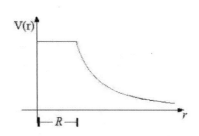

(D)
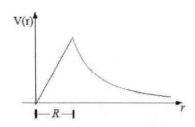

(E)
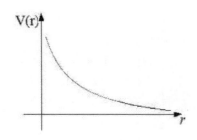

47. If a thin infinitely long straight wire of uniform linear charge density λ produces an electric field **E** at a radial distance r from the axis of wire, the magnitude of **E** is

(A) $E = \dfrac{\lambda}{4\pi r \varepsilon_0}$

(B) $E = \dfrac{\lambda}{2\pi r \varepsilon_0}$

(C) $E = \dfrac{2\lambda}{\pi r \varepsilon_0}$

(D) $E = \dfrac{\varepsilon_0 \lambda}{2\pi r}$

(E) $E = \dfrac{\lambda}{\pi r \varepsilon_0}$

48. Consider a non-conducting charged sphere of radius R having charge Q distributed uniformly in the spherical volume. The electric field E due to such a sphere as a function of r from its centre is represented graphically by

(A)
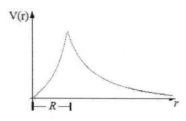

(B)
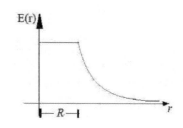

(C)

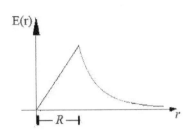

(D)

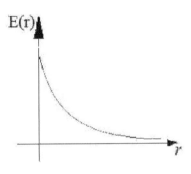

(E)

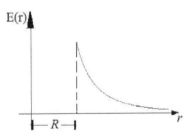

49. As shown in the figure, charges Q and $4Q$ are uniformly distributed in two non-conducting solid spheres S_1 and S_2 of radii r and $4r$ respectively. If magnitude of the electric fields at points A and B at a distance $2r$ from the centres O_1 and O_2 of spheres S_1 and S_2 respectively are E_1 and E_2 respectively, then

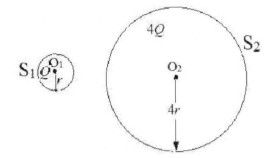

(A) $E_1 > E_2$

(B) $E_1 < E_2$

(C) $E_1 = E_2$

(D) E_1 could be greater than or less than E_2

(E) None of the above

50. Consider a hollow cylinder having a charge Q coulomb within it. If the electric flux through each of the plane end surfaces A and B be φ_p, the flux φ_c linked with the curved surface of the cylinder will be

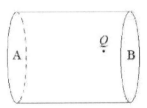

(A) $\dfrac{Q}{\varepsilon_0} - \varphi_p$

(B) $\dfrac{Q}{\varepsilon_0} - 2\varphi_p$

(C) $\dfrac{Q}{2\varepsilon_0} - \varphi_p$

(D) $2\varphi_p + \dfrac{Q}{\varepsilon_0}$

(E) $\varphi_p + \dfrac{Q}{\varepsilon_0}$

51. Consider a very large non-conducting sheet having a positive uniform surface charge density σ. The electric field **E** at a point at small distance r in front of the sheet is

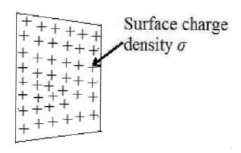

(A) is perpendicular to the sheet and has the magnitude $(\frac{\sigma}{\varepsilon_0})$

(B) is perpendicular to the sheet and has the magnitude $(\frac{\sigma}{2\varepsilon_0 r})$

(C) is parallel to the sheet and has the magnitude $(\frac{2\sigma}{\varepsilon_0 r})$

(D) is perpendicular to the sheet and has the magnitude $(\frac{\sigma}{2\varepsilon_0})$

(E) is parallel to the sheet and has the magnitude $(\frac{\sigma}{2\varepsilon_0})$

52. A Gaussian surface

(A) can pass through a continuous charge distribution but cannot pass through a discrete charge

(B) cannot pass through a continuous charge distribution but can pass through a discrete charge

(C) can pass through both a discrete charge as well as continuous charge distribution

(D) cannot pass through either a continuous charge distribution or through a discrete charge

(E) none of the above

53. The figure given below shows three infinitely large charge sheets A, B and C having charge densities as shown. The electric field at point P is

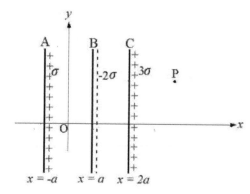

(A) $\left(\frac{\sigma}{2\varepsilon_0}\right)$ along negative x-axis

(B) $\left(\frac{\sigma}{\varepsilon_0}\right)$ along positive x-axis

(C) $(\frac{\sigma}{2\varepsilon_0})$ along positive x-axis

(D) $\left(\frac{2\sigma}{\varepsilon_0}\right)$ along negative x-axis

(E) $(\frac{\sigma}{2\varepsilon_0})$ along negative y-axis

54. For a given closed surface the Gauss's law is stated as $\oint E.dA = 0$. From this we can conclude that

(A) the flux is only going into the surface

(B) Electric field **E** is necessarily zero on the surface

(C) The flux is only going out of the surface

(D) Electric field **E** is necessarily parallel to the surface at every point

(E) The charge enclosed by the surface is zero

1.5 Fields and potentials of other charge distributions

Questions 55-56

A non-conducting semi-circular ring of radius r has a positive charge distributed uniformly over it and the charge per unit length is λ.

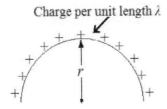

55. The magnitude of electric field at the center of the ring is

    (A) $\dfrac{\lambda}{2\pi\varepsilon_0 r}$

    (B) $\dfrac{\lambda}{\pi\varepsilon_0 r}$

    (C) $\dfrac{2\lambda}{\pi\varepsilon_0 r}$

    (D) $\dfrac{\lambda}{4\pi\varepsilon_0 r}$

    (E) $\dfrac{\lambda\pi}{\varepsilon_0 r}$

56. The electric potential at the centre of this ring is

    (A) $\dfrac{\lambda}{\varepsilon_0}$

    (B) $\dfrac{\lambda}{4\varepsilon_0}$

    (C) $\dfrac{4\lambda}{\varepsilon_0}$

    (D) $\lambda\,\varepsilon_0$

    (E) $2\lambda\,\varepsilon_0$

57. A circular ring made of non-conducting material of radius R has a positive charge +Q uniformly distributed along one-half of its circumference and a negative charge of –Q uniformly distributed along the other half of the circumference. It is placed in the x-y plane with its center O coinciding with the origin as shown.

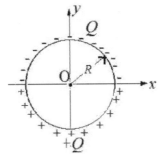

(A) The electric field at O is $\mathbf{E} = \dfrac{Q}{\pi^2\varepsilon_0 R^2}\,\mathbf{j}$

(B) The electric field is zero at O

(C) The electric field at O is $\mathbf{E} = \dfrac{2Q}{\pi\varepsilon_0 R^2}\,\mathbf{j}$

(D) The electric potential at O is $V = \dfrac{Q}{4\pi\varepsilon_0 R}$

(E) The electric potential is not zero anywhere inside the ring

58. A charge Q is uniformly distributed over a long rod AB of length L as shown in the figure. The electric potential V_P at the point P lying at distance L from the end B varies as

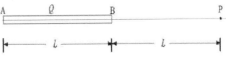

(A) L

(B) L^2

(C) $\dfrac{1}{L^2}$

(D) $\dfrac{1}{L}$

(E) e^{-L}

Questions 59-60

Consider a circular ring made of non-conducting material having radius r. It has a positive charge q distributed uniformly over it. Let P be a point on the axis (a line perpendicular to the plane of the ring and passing through the centre O of the ring) at a distance x from O.

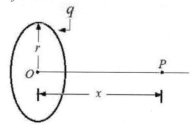

59. The magnitude of electric field at P is

   (A) $\dfrac{1}{4\pi\varepsilon_0}\dfrac{qx}{(r^2+x^2)^{1/2}}$

   (B) $\dfrac{1}{4\pi\varepsilon_0}\dfrac{qx}{(r^2+x^2)^{3/2}}$

   (C) $\dfrac{1}{4\pi\varepsilon_0}\dfrac{2qx}{(r^2+x^2)^{3/2}}$

   (D) $\dfrac{1}{8\pi\varepsilon_0}\dfrac{qx}{(r^2+x^2)^{3/2}}$

   (E) $\dfrac{1}{4\pi\varepsilon_0}\dfrac{qx^2}{(r^2+x^2)^{5/2}}$

60. The electric potential V at P is

   (A) $\dfrac{1}{4\pi\varepsilon_0}\dfrac{qx}{(r^2+x^2)^{1/2}}$

   (B) $\dfrac{1}{4\pi\varepsilon_0}\dfrac{q}{x(r^2+x^2)^{1/2}}$

   (C) $\dfrac{1}{8\pi\varepsilon_0}\dfrac{q}{(r^2+x^2)^{1/2}}$

   (D) $\dfrac{1}{4\pi\varepsilon_0}\dfrac{2q}{(r^2+x^2)^{1/2}}$

   (E) $\dfrac{1}{4\pi\varepsilon_0}\dfrac{q}{(r^2+x^2)^{1/2}}$

61. As shown below, two identical thin rings A and B, each of radius R are coaxially placed at a distance R apart. If Q_1 and Q_2 are respectively the charges uniformly spread on the two rings, the electric potential at the center O_2 of B is

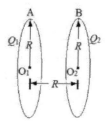

   (A) $\left(\dfrac{1}{4\pi\varepsilon_0}\right)\left(\dfrac{Q_2}{\sqrt{2}R}+\dfrac{Q_1}{R}\right)$

   (B) $\left(\dfrac{1}{4\pi\varepsilon_0}\right)\left(\dfrac{Q_1}{\sqrt{2}R}+\dfrac{Q_2}{R}\right)$

   (C) $\left(\dfrac{1}{4\pi\varepsilon_0}\right)\left(\dfrac{Q_1}{R}+\dfrac{Q_2}{R}\right)$

   (D) $\left(\dfrac{1}{4\pi\varepsilon_0}\right)\left(\dfrac{Q_1}{\sqrt{2}R}+\dfrac{2Q_2}{R}\right)$

   (E) $\left(\dfrac{1}{4\pi\varepsilon_0}\right)\left(\dfrac{Q_1}{2R}+\dfrac{Q_2}{R}\right)$

62. An electric field **E** exists in space produced by a total charge Q distributed non-uniformly on the circumference of a non-conducting circular ring of radius R. The value of integral $\int_{x=\infty}^{x=0}-\mathbf{E}\cdot d\mathbf{x}$ ($x=0$ denotes the center O of the ring) is

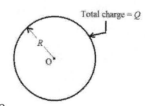

   (A) $\dfrac{Q}{\varepsilon_0 R}$

   (B) $\dfrac{2Q}{R}$

   (C) $\dfrac{Q}{\varepsilon_0 R^2}$

   (D) $\dfrac{Q}{\pi\varepsilon_0 R}$

(E) $\dfrac{Q}{4\pi\varepsilon_0 R}$

UNIT 2: Conductors, Capacitors, Dielectrics

2.1 Electrostatics with Conductors

63. Two uncharged metallic spheres A and B are supported on insulating stands and made to touch each other as shown in the figure given below.

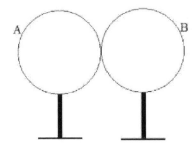

A glass rod rubbed with silk is brought close to sphere A. When the glass rod is subsequently removed and the spheres are separated far apart

(A) the two spheres will remain uncharged

(B) there will be negative charge distributed uniformly on A and positive charge distributed uniformly on B

(C) there will be negative charge on A and positive charge on B facing each other

(D) there will be negative charge on A and no change on B

(E) there will be positive charge distributed uniformly on A and negative charge distributed uniformly on B

64. Suppose an extra charge Q is given to a neutral metallic conductor.

Choose the correct choice among the following

(A) Electric field will be non-zero within the conductor. Electric field is normal to the surface S of conductor at every point. Electric potential has a constant value within the conductor and equals the electric potential at the surface S.

(B) Electric field will be non-zero within the conductor. Extra charge given to the uncharged conductor cannot reside on the surface S of the conductor. Electric field is normal to S at every point of S. Electric potential has a constant value within the conductor and equals the electric potential at the surface S.

(C) Electric field will be zero within the conductor. Extra charge given the uncharged conductor will reside on the surface S of the conductor. Electric field will be normal to S at every point of S. Electric potential will have a constant value within the conductor, which will be equal to the electric potential at the surface S.

(D Electric field will be zero within the conductor. Extra charge given to an uncharged conductor must reside on the surface S of the conductor. Electric field is tangential to S at every point of S. Electric potential has a constant value within the conductor and equals the electric potential at the surface S.

(E) Electric field will be non-zero within the conductor. Extra charge given to an uncharged conductor must reside on the surface of the conductor. Electric field is normal to S at every point of S. Electric potential has a constant value within the conductor, which is different from electric potential at the surface S.

Questions 65-66

Consider a charged conducting sphere of radius R having a charge Q. Let r be the distance from the center O of the sphere.

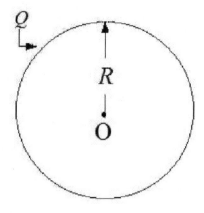

65. The variation of magnitude of electric field with the distance r from the centre, is best described by

(A)

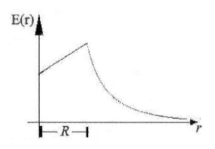

(B)

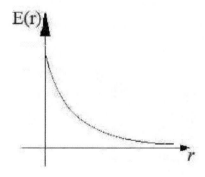

(C)

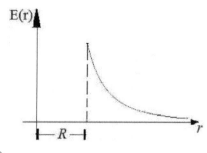

(D)

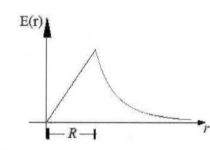

(E)

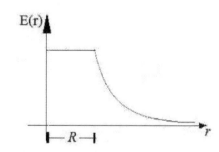

66. The variation of magnitude of electric potential with the distance r from the centre, is best described by

(A)

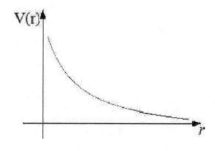

(B)
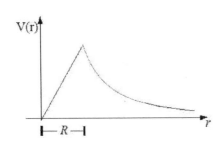

(C)
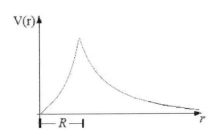

(D)
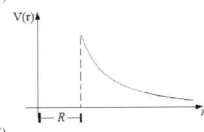

(E)
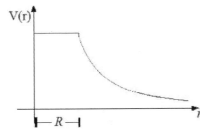

67. If a charged conducting spherical shell of radius 6 cm has potential 12 V at a point A distant 2 cm from its centre O, then the potential at O and at a point B distant 10 cm from O are, respectively

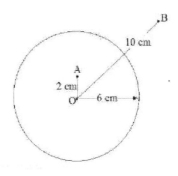

(A) 12 V, 12 V
(B) zero, 7.2 V
(C) 12 V, 7.2 V
(D) 12 V, 6.0 V
(E) zero, 12 V

68. Suppose a solid conductor having charge Q has a cavity as shown below. If a charge q is kept inside the cavity, the charge at the inner and outer surface respectively of the conductor are

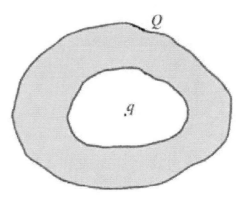

(A) $(q, Q - q)$
(B) $(q, Q + q)$
(C) $(-q, Q + q)$
(D) $(-q, Q - q)$
(E) (q, Q)

69. A metallic spherical shell has a positive point charge q kept inside a spherical cavity. The diagram which correctly represents the electric field lines is

(A)
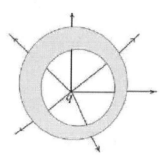

(B)
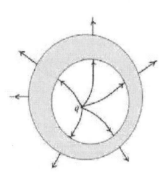

(C)
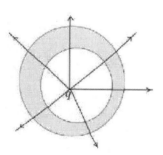

(D)
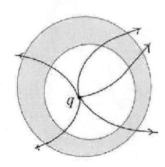

(E)
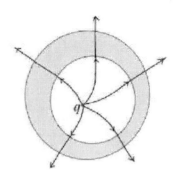

70. A uniform electric field **E** exists in space. When a metallic solid sphere is placed in the electric field, the electric field lines follow the path shown in figure as

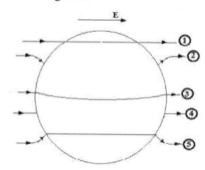

(A) 1
(B) 2
(C) 3
(D) 4
(E) 5

71. Suppose the surface charge density on the surface of a charged conductor is $+\sigma$. Then the magnitude of electric field at a point P on the surface is

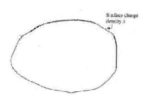

(A) $\dfrac{\varepsilon_0 \sigma}{2}$

(B) $\dfrac{\sigma}{2\varepsilon_0}$

(C) $\dfrac{2\sigma}{\varepsilon_0}$

(D) $\dfrac{\sigma}{\varepsilon_0}$

(E) $\varepsilon_0 \sigma$

72. A conductor with a non-spherical shape is shown below.

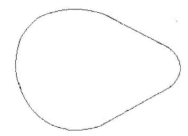

Suppose some positive charge is given to this conductor. Which among the following diagrams best represents the distribution of charge and the electric field lines on the conductor?

(A)

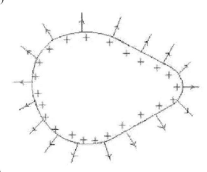

(B)

(C)

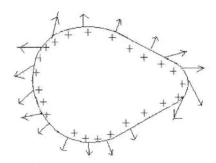

(D)

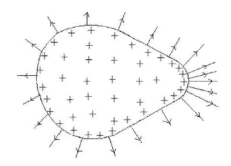

(E)

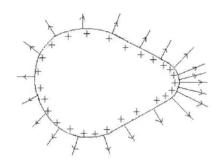

73. Consider a large, plane, uniformly charged thin conducting sheet S in a vertical plane having a surface charge density σ. As shown in the figure given below, a small charged ball B having charge $+Q$ hangs by a silk thread from a wooden peg near the sheet. If the thread makes an angle θ with S, we have

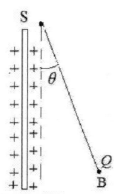

Side view of the conducting sheet

(A) $\sin \theta = \dfrac{\sigma Q}{\varepsilon_0 mg}$

(B) $\tan \theta = \dfrac{\sigma Q}{\varepsilon_0 mg}$

(C) $\cos \theta = \dfrac{\sigma Q}{\varepsilon_0 mg}$

(D) $\tan \theta = \dfrac{Q}{\varepsilon_0 mg\, \sigma}$

(E) $\tan \theta = \varepsilon_0 \dfrac{\sigma Q}{mg}$

74. Consider two concentric conducting thin spherical shells A and B having radii a and b ($a < b$) carrying charges q and $-Q$ ($|q| < |Q|$) respectively. Let r be the distance from the common center O of the shells. The graph of electric field as a function of r is best represented by

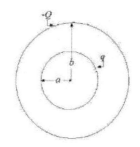

(A)

(B)

(C)

(D)

(E)

2.2 Capacitors

75. In the figure given below, three capacitors of capacity equal to $C_1 = C$, $C_2 = 2C$, $C_3 = 3C$ are conducted to a battery as shown in the figure. The ratio of the charges on C_1 and C_3 is

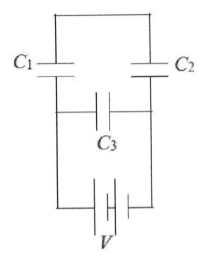

(A) $\dfrac{2}{3}$

(B) $\dfrac{2}{7}$

(C) $\dfrac{4}{9}$

(D) $\dfrac{2}{9}$

(E) $\dfrac{9}{2}$

76. Four identical metal plates having area A are located in air at equal distance d from one another as shown. The capacitance of the system between points P and Q is

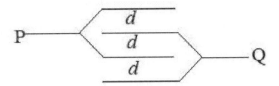

(A) $3\,\varepsilon_0 \dfrac{A}{d}$

(B) $6\,\varepsilon_0 \dfrac{A}{d}$

(C) $4\,\varepsilon_0 \dfrac{A}{d}$

(D) $\varepsilon_0 \dfrac{A}{3d}$

(E) $2\,\dfrac{A}{\varepsilon_0 d}$

77. Suppose we have three identical capacitors. If these capacitors are placed in series, the equivalent capacitance of the series combination is C. If now the three capacitors are place in parallel, the equivalent capacitance of parallel combination will be

(A) $3\,C$

(B) $\dfrac{C}{9}$

(C) $9\,C$

(D) $\dfrac{C}{3}$

(E) $6\,C$

78. Consider two large plane conducting plates A and B each having area A. Suppose they are given positive charges Q and q ($q < Q$) respectively and brought close together to form a parallel plate capacitor. If the capacity of the capacitor so formed is C when the separation between A and B is x, the potential difference V between A and B will be

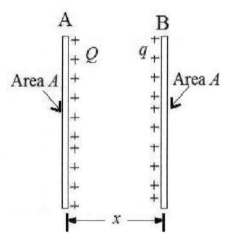

(A) $\dfrac{(Q-q)}{C}$

(B) $\dfrac{(Q-q)}{2C}$

(C) $\dfrac{Q}{2C}$

(D) $\dfrac{(Q+q)}{2C}$

(E) $\dfrac{(Q+q)}{C}$

79. Suppose the intervening medium between the plates of a parallel plate capacitor is vacuum. If a uniform electric field of magnitude E exists in the space between the plates, the energy stored per unit volume in the inner region between the plates is

(A) $\dfrac{1}{3}(\varepsilon_0 E^2)$

(B) $\dfrac{1}{2}(\varepsilon_0 E^2)$

(C) $\dfrac{1}{2}(E^2/\varepsilon_0)$

(D) $\varepsilon_0 E$

(E) $\varepsilon_0 E^2$

80. A parallel plate capacitor is charged and the charging battery is then disconnected. Suppose the plates of the capacitor are moved farther apart by means of insulating handles. Choose the correct answer regarding the effect of increasing the plate separation on charge of the capacitor, capacitance of the capacitor, potential difference between the plates, and the energy stored in the capacitor.

(A) Charge remains constant, capacitance increases, potential difference increases and energy increases

(B) Charge remains constant, capacitance decreases, potential difference increases and energy increases

(C) Charge remains constant, capacitance decreases, potential difference increases and energy decreases

(D) Charge decreases, capacitance decreases, potential difference increases and energy increases

(E) Charge remains constant, capacitance decreases, potential difference decreases and energy increases

81. Six capacitors, each of capacitance value C are connected to form a hexagon as shown in the

figure. The ratio of capacitance between P and Q to the capacitance between P and R is

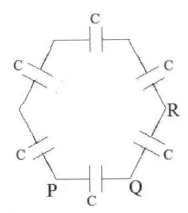

(A) $\dfrac{3}{5}$

(B) $\dfrac{8}{3}$

(C) $\dfrac{9}{5}$

(D) $\dfrac{9}{4}$

(E) $\dfrac{8}{5}$

82. If a parallel combination of 5 capacitors, each of value C_1, when charged by a source of potential difference V has the same value of total stored energy as a series combination of 7 capacitors, each of value C_2 when charged by another source of potential difference $4V$, the ratio (C_1 / C_2) is

(A) $\dfrac{8}{35}$

(B) $\dfrac{11}{35}$

(C) $\dfrac{16}{21}$

(D) $\dfrac{16}{35}$

(E) $\dfrac{4}{35}$

83. Three capacitors each of 6 μF are to be connected to obtain an effective capacitance is 9 μF. The combination that will achieve the desired result is

(A)

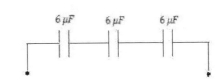

(B)

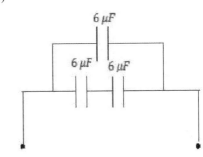

(C)

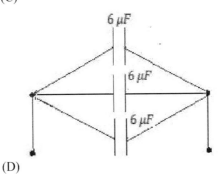

(D)

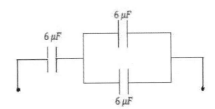

(E) None of the above.

34

84. A capacitor having capacitor C is charged is charged to potential V when the switch S is turned to position 1 as shown. When S is turned to position 2, does the stored energy increase or decrease and by how much percentage?

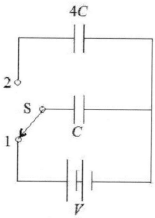

(A) Decreases, 80%
(B) Increases, 40%
(C) Decreases, 25%
(D) Decreases, 75%
(E) Increases, 10%

85. Consider two co-axial long cylindrical conductors having length L. Suppose a and b are the radii of inner and outer cylinders respectively and $b \ll L$. Suppose charge $+Q$ is given to the inner cylinder. The capacitance of the cylindrical capacitor so formed is

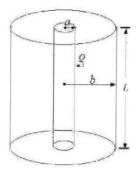

(A) $2\pi\varepsilon_0 L \ln\frac{b}{a}$

(B) $\dfrac{2\pi\varepsilon_0 L}{\ln\frac{b}{a}}$

(C) $\dfrac{L}{2\pi\varepsilon_0 \ln\frac{b}{a}}$

(D) $\dfrac{4\pi\varepsilon_0 L}{\ln\frac{b}{a}}$

(E) $4\pi\varepsilon_0 L \ln\frac{b}{a}$

2.3 Dielectrics

86. he figure given below shows a parallel plate capacitor whose positive plate is at $x = 0$ and negative plate is at $x = 5d$. Two dielectric slabs S_1 and S_2, each of thickness d are inserted between plates with gap d between the slabs. The dielectric constants of the dielectric materials of S_1 and S_2 are K_1 and K_2 ($K_1 > K_2$) respectively.

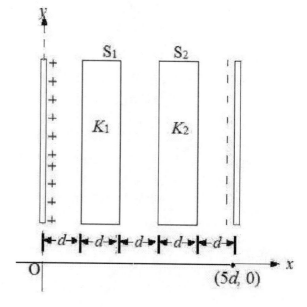

As x goes from 0 to $5d$, the variation of electric field is correctly shown by

(A)
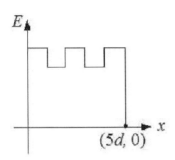

(B)
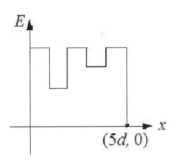

(C)
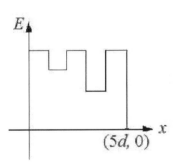

(D)
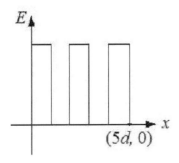

(E)
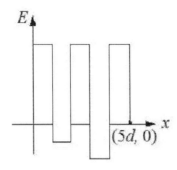

Questions 87-88

A parallel plate capacitor is having plate area A and separation between the plates d. When there is vacuum between the plates, its capacitance is C_0. This capacitor is charged to a potential difference V_0.

87. When a slab of material of dielectric constant K having the same plate area as that of parallel-plate capacitor but having a thickness t is inserted parallel to the plates of the parallel-plate capacitor, the capacitance value changes to C. The ratio $\frac{C}{C_0}$ is

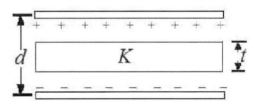

(A) $1 + \frac{t}{dK}(K-1)$

(B) $\frac{t}{dK}(1-K)$

(C) $1 - \frac{t}{dK}(K-1)$

(D) $1 + \frac{t\,d}{K}(1-K)$

(E) $1 + t\,dK\,(1-K)$

88. If, instead of a slab of material of dielectric constant K, a metal plate having a thickness t is inserted parallel to the plates of the parallel-plate capacitor, the capacitance will be

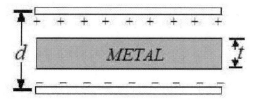

(A) $K t C_0$
(B) $K t d C_0$
(C) C_0 / t
(D) $\dfrac{C_0 t}{K d}$
(E) $\dfrac{C_0}{\left(1 - \dfrac{t}{d}\right)}$

89. A parallel plate capacitor having plate area A and separation between the plates d has capacitance C_0 when the intervening medium between the plates is vacuum. Two thin dielectric slabs of thickness d having dielectric constants K_1 and K_2 are inserted between plates of the capacitor, as shown in the figure. If C be the capacitance of the arrangement, the ratio $\left(\dfrac{C}{C_0}\right)$ will be

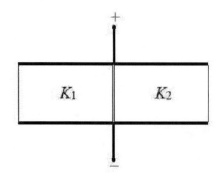

(A) $2(K_1 + K_2)$
(B) $(K_1 + K_2)$
(C) $\dfrac{1}{2} (K_1 / K_2)$
(D) $\dfrac{1}{2} (K_1 + K_2)$
(E) K_1 / K_2

90. A parallel plate capacitor with vacuum between the plates is charged by connecting it to a battery. Let the quantities capacitance, charge, voltage, electric field and energy associated with the charged capacitor be denoted as C_0, Q_0, V_0, E_0 and U_0 respectively. With the battery still connected, suppose a dielectric slab having dielectric constants K is introduced to fill the space between the plates. Suppose the corresponding quantities are now given by C_K, Q_K, V_K, E_K and U_K respectively. Choose the correct answer among the following.

(A) $C_K > C_0, Q_K > Q_0, V_K = V_0, E_K = E_0$ and $U_K > U_0$
(B) $C_K < C_0, Q_K < Q_0, V_K = V_0, E_K = E_0$ and $U_K < U_0$
(C) $C_K < C_0, Q_K < Q_0, V_K < V_0, E_K < E_0$ and $U_K < U_0$
(D) $C_K > C_0, Q_K > Q_0, V_K = V_0, E_K = E_0$ and $U_K < U_0$
(E) $C_K > C_0, Q_K = Q_0, V_K = V_0, E_K = E_0$ and $U_K > U_0$

B. FREE-RESPONSE QUESTIONS

91. In figure given below, two small balls of identical mass m and having identical charge q are suspended from the same point with the help of two insulating strings of length l. They diverge due to electrostatic force of repulsion and reach equilibrium with each making an angle θ with the vertical. θ is so small that $\tan \theta$ can be replaced by $\sin \theta$.

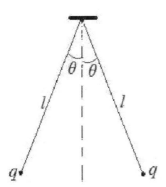

(a) When equilibrium is reached, how much is the

    (i) separation between the two balls?

    (ii) the tension in each string

(b) 35 yrs iit/130 4 bit/43 3 Suppose that the charge begins to leak from both the spheres at a constant rate. Prove that the velocity with which the charges approach each other at any instant is inversely proportional to the square root of separation between the balls at that instant.

(c) 41 yrs iit/97 3 Imagine now that the arrangement of the two charged balls is taken in space where there is no gravitational effect. What will be the values of θ and how much will there be the tension in each string?

92. wo charges $-q$ and $2q$ are kept at points A $(0, 3a)$ and B $(0, -3a)$ respectively of an x-y coordinate system.

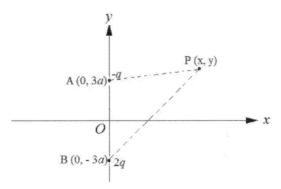

(a) What is the electric potential at a point P (x, y) in the x-y plane due to the two charges?

(b) Find the equation of the curve on all points of which the electric potential due to the two charges is zero.

(c) Find the coordinates of all points having zero electric potential on

    (i) the x-axis

    (ii) the y- axis

(d) Find the electric field at the point nearest to the origin O among those in part (c) above.

93. Consider three charges each equal to q at the vertices of an equilateral triangle ABC of side a. A fourth charge Q is placed at the centroid O of the triangle as shown in the figure given below.

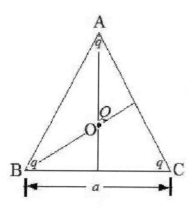

(a) What is the net force on the charge q at A?
(b) What should be the ratio (Q/q) so that the charges at the vertices remain stationary?
(c) If $(Q/q) = -1$, will the charges at the vertices move towards O or away from it?
(d) For the value of Q for which the charges at the vertices are stationary, find the potential energy of the system?

94. A long non-conducting cylinder of radius R has a non-uniform charge distribution with volume charge density
$\rho(r) = \rho_0 r^2$
where ρ_0 is a constant and r is the distance measured from the axis of the cylinder. The charge density is zero outside the cylinder.
(a) The charge contained in length l of a cylinder of radius $r > R$ is
(b) The electric field for $r < R$ is
(c) The electric field for $r > R$ is

(d) Substitute $r = R$ in expressions for electric field obtained in parts (b) and (c). Interpret the result so obtained.

95. Charge Q is non-uniformly distributed in a non-conducting solid sphere of radius R. The variation of the charge density $\rho(r)$ with the radial distance r from the center O of the sphere is as per the diagram given below.

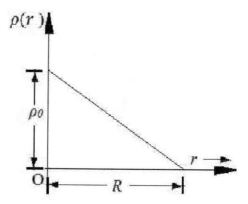

where ρ_0 is a constant.
(a) The electric field for $r > R$ is
(b) The electric field for $r < R$ is
(c) Express ρ_0 in terms of Q.
(d) Calculate the electric potential at the center $(r = 0)$ of the sphere.
(e) Verify the result in part (d) by summing up the contribution of spherical shells into which the sphere can be divided.

96. Consider a non-conducting charged disc of radius R, having uniform surface charge density σ. Let P be a point a point on the axis (i.e., on the line perpendicular to the surface of the disk passing through the center O of the disk) at a distance x from O.

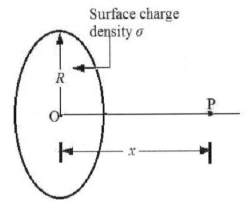

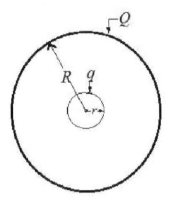

(a) Calculate the electric potential at P due to the charged disk.
(b) Calculate the electric field at P due to the charged disk.
(c) Also calculate electric field at P by differentiating the value of electric potential at P obtained in part (a) and check whether you get the same answer as in part (b) above.
(d) Make suitable approximation and check whether the result for electric field obtained above is consistent with
   (i) the electric field due to a point charge
   (ii) the electric field near an infinitely large non-conducting sheet of charge.

97. Consider a large spherical conducting shell of radius R having charge Q. Concentric with the shell is a small conducting sphere of radius r, carrying some positive charge q.

(a) What is the electric potential at the surface of the spherical shell?
(b) Calculate the electric potential at the surface of the small sphere.
(c) Suppose the inner sphere is connected to the outer shell with a metallic wire. Show that, irrespective of charge on the shell, any charge given to the sphere will flow to the shell.
(d) The above result that more and more charge can be transferred to the outer metallic shell does not mean that unlimited charge can accumulate on the shell. Can you think why?
(e) If the breakdown field of air is 3×10^8 V/m, estimate the value of electric potential to which a metallic shell of radius 0.5 m can be raised before dielectric breakdown of the air occurs

98. In figure given below, the four capacitors having capacitances $C_1 = 3\ \mu F$, $C_2 = 6\ \mu F$, $C_3 = 2\ \mu F$ and $C_4 = 4\ \mu F$ are joined to a battery of 12 V.

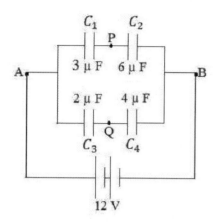

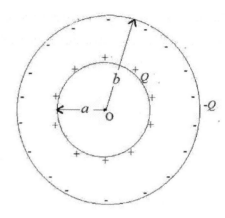

(a) What is the equivalent capacitance of the network between the points A and B.

(b) How much charge is supplied by the battery?

(c) Find the charges on each of the four capacitors.

(d) Suppose now that a fifth capacitor $C_5 = 5\ \mu F$ is joined between the points P and Q,

    (i) What will be the charge on C_5?

    (ii) What will be the new values of the quantities calculated in (a), (b) and (c)?

99. Consider a hollow (or solid) spherical conductor together surrounded by a concentric spherical shell. Let the radii of two conductors be a and b respectively ($b > a$). Positive charge Q is given to the inner conductor and $-Q$ is given to the outer shell.

(a) Find the potential difference between the inner and the outer spheres.

(b) What is the capacitance of the spherical capacitor formed by the above pair of spherical conductors?

(c) Consider a single spherical conductor of radius a, which is given charge Q. Calculate the capacitance (= ratio of charge to the potential) of such an isolated conductor.

(d) Think how you can make suitable approximations in the result about capacitance of the spherical capacitor discussed in part (b) above to arrive at the

    (i) Capacitance of a single spherical conductor

    (ii) Capacitance of a parallel plate capacitor

(e) Compare the radius of an isolated spherical conductor that has a capacitance of 1 Farad to the radius of earth. What conclusion do you derive from this ratio?

100. Consider two parallel plate capacitors A and B. When there is vacuum between the plates, the capacitance of each of them is C0.

(a) A slab of dielectric of dielectric constant K is inserted to exactly fill the space between the plates of capacitor A, which is then charged by joining to a battery of voltage V. How much is the electrostatic energy stored in the capacitor?

(b) The battery is disconnected and then the dielectric slab is removed. Does the stored energy increase or decrease? Why?

(c) With the battery remaining disconnected, the same slab of dielectric is inserted to exactly fill the space between the plates of capacitor B. The two capacitors A (without the slab) and B (with the slab) are now joined together in parallel. Calculate the equivalent capacitance of the combination of two capacitors.

(d) What is the electrostatic energy stored in the combination of two capacitors in part (c)?

PART II: ANSWERS

ANSWER KEY

1. C
2. B
3. B
4. D
5. E
6. C
7. B
8. B
9. E
10. D
11. D
12. B
13. C
14. B
15. A
16. C
17. E
18. C
19. B
20. A
21. B
22. C
23. C
24. C
25. B
26. D
27. C
28. A
29. B
30. C
31. D
32. B
33. C
34. B
35. B
36. A
37. A
38. C
39. B
40. C
41. C
42. B
43. B
44. A
45. E
46. C
47. B
48. C
49. A
50. B
51. D
52. A
53. B
54. E
55. A
56. B
57. A
58. D
59. B
60. E
61. B
62. E
63. B
64. C
65. C
66. E
67. C
68. C
69. B
70. B
71. D
72. D
73. B
74. D
75. D
76. A
77. C
78. B
79. B
80. B
81. E
82. D
83. B
84. A
85. B
86. B
87. C
88. E
89. D
90. A

ANSWERS AND EXPLANATIONS

A. MULTIPLE-CHOICE QUESTIONS

UNIT 1: Electrostatics

1.1 Charge and Coulomb's Law

1. **C**

   Let $q = n\,e$

   Now,

   $$F = k\frac{q^2}{r^2}$$

   $$q^2 = n^2 e^2 = F\frac{r^2}{k} = 4\pi\varepsilon_0 F\, r^2$$

   $$n = \sqrt{\frac{4\pi\varepsilon_0 F\, r^2}{e^2}}$$

   We find that n is directly proportional to $\sqrt{F}$.

2. **B**

   According to the principle of superposition, when several forces act on any charge due to a number of other charges, the resultant force on the charge is the vector sum of all the individual forces on that charge due to the other charges, taken one at a time. The individual forces are unaffected due to the presence of other charges.

3. **B**

   Since both the electrostatic force of repulsion as well as the gravitational force of attraction act along the line joining the two particles, the magnitudes of these two forces should be equal.

   $$k\frac{q^2}{r^2} = G\frac{m^2}{r^2}$$

   $$\frac{q}{m} = \sqrt{\frac{G}{k}} = \sqrt{4\pi\varepsilon_0 G}$$

4. **D**

   Clearly, the charge Q, being at the centre of two identical charges, is in equilibrium.

   Distance between q and $Q = a$

   Distance between q and $q = 2a$

For equilibrium of q at x = + a, the net force on it due to two other charges should be zero.

$$k\frac{qQ}{a^2} + k\frac{q^2}{(2a)^2} = 0$$

$$4qQ + q^2 = 0$$

$$Q = -\frac{q}{4}$$

5. **E**

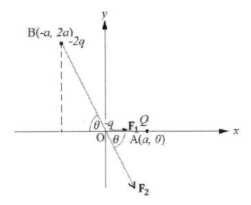

$OB = \sqrt{a^2 + 4a^2} = \sqrt{5}a$

Let $\mathbf{F_1}$ and $\mathbf{F_2}$ be the vectors representing the forces on $-q$ at O due to charges Q at A and $-2q$ at B respectively.

The x-component of force of attraction $\mathbf{F_1}$

$$F_{1x} = k\frac{qQ}{a^2}\cos 0^0$$

The x-component of force of repulsion $\mathbf{F_2}$

$$F_{2x} = k\frac{(q)(2q)}{(\sqrt{5}a)^2}\cos\theta$$

The x-component of net force $\mathbf{F}$ on $-q$ at O

$$F_x = F_{1x}\mathbf{i} + F_{2x}\mathbf{i}$$

$$= (k\frac{qQ}{a^2} + k\frac{2q^2}{5a^2}\cos\theta)\mathbf{i}$$

$$= (k\frac{qQ}{a^2} + k\frac{2q^2}{5a^2}\frac{a}{\sqrt{5}a})\mathbf{i}$$

$$= kq\left(\frac{Q}{a^2} + \frac{2q}{5\sqrt{5}a^2}\right)\mathbf{i}$$

$$F_x \propto \left(\frac{Q}{a^2} + \frac{2q}{5\sqrt{5}a^2}\right)$$

46

 6. C

The force on $-q$ at O due to charges Q at A is F_1. It is a force of attraction along OA i.e., along x-axis. So, its y-component is zero.

The force on $-q$ at O due to charges $-2q$ at B is F_2. It does not depend on Q.

Hence, the net force on $-q$ at O, being the vector sum of F_1 and F_2, is independent of Q.

 7. B

$$F_1 = k \frac{qQ}{(a)^2} \cos 0° \, \mathbf{i}$$

$$F_2 = k \frac{(q)(2q)}{(\sqrt{5}a)^2} \cos \theta \, \mathbf{i} - k \frac{(q)(2q)}{(\sqrt{5}a)^2} \sin \theta \, \mathbf{j}$$

Net force on on $-q$ at O

$$F = F_1 + F_2$$

$$= (k \frac{qQ}{a^2} + k \frac{2q^2}{5a^2} \cos \theta) \mathbf{i} - k \frac{2q^2}{5a^2} \sin \theta \, \mathbf{j}$$

$$= (k \frac{qQ}{a^2} + k \frac{2q^2}{5a^2} \frac{a}{\sqrt{5}a}) \mathbf{i} - k \frac{2q^2}{5a^2} \frac{2a}{\sqrt{5}a} \mathbf{j}$$

$$F = \frac{q}{4\pi\varepsilon_0} \left(\frac{Q}{a^2} + \frac{2q}{5\sqrt{5}\,a^2} \right) \mathbf{i} - \frac{1}{\pi\varepsilon_0} \frac{q^2}{5\sqrt{5}\,a^2} \mathbf{j}$$

8. **B**

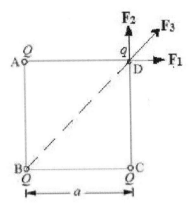

The magnitude of force F_1 on q due to charge Q at A is

$$F_1 = k\frac{Qq}{a^2}$$

F_1 is along AD.

The force F_2 on q due to charge Q at C will also have the magnitude

$$F_2 = k\frac{Qq}{a^2} = F_1$$

F_2 will be along CD.

Now BD = $\sqrt{2}\,a$.

So, the magnitude of force F_3 on q due to charge at B is

$$F_3 = k\frac{Qq}{(\sqrt{2}\,a)^2} = k\frac{Qq}{2a^2}$$

The direction of F_3 is along BD.

Note that F_1 and F_2 have equal magnitude and are at right angle to each other. So, their resultant F' will be along the bisector of angle between them and hence along DR, which coincides with the direction of F_3. Also, the magnitude of F' is

$$F' = \sqrt{(F_1^2 + F_2^2)} = \sqrt{2}\,k\frac{Qq}{a^2}$$

Hence, the resultant force on charge q at D due to the three charges at A, B and C will have the magnitude

$$F_R = F' + F_3 = \sqrt{2}\,k\frac{Qq}{a^2} + k\frac{Qq}{2a^2}$$

$$F_R = k\frac{Qq}{a^2}\left(\sqrt{2} + \frac{1}{2}\right)$$

Hence,

$$\frac{F_R}{F_1} = \sqrt{2} + \frac{1}{2}$$

9. **E**

Let the original charge on each of the spheres of negligible size be q and the distance between their centers be r.

The magnitude of the electrostatic force of repulsion on each is given by

$$F = k\frac{q^2}{r^2} = 10^{-5}\text{ N}$$

When identical uncharged sphere C touches A, the charges redistribute and, by symmetry, both A and C carry a charge $q/2$. When C is placed at mid-point between A and B, the force on C due to charge $q/2$ at A is (say) $\mathbf{F_1}$. $\mathbf{F_1}$ is directed towards AC and its magnitude is

$$F_1 = k\frac{(q/2)(q/2)}{\left(\frac{r}{2}\right)^2} = F$$

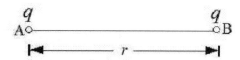

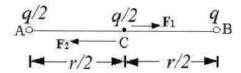

Magnitude of force between B and C becomes

$$F_2 = k\frac{(q)(q/2)}{\left(\frac{r}{2}\right)^2} = 2F$$

$\mathbf{F_2}$ is towards BC.

Thus the net force on the sphere C is

$$\mathbf{F_C} = \mathbf{F_1} + \mathbf{F_2}$$

The direction of $\mathbf{F_C}$ is from C towards A and its magnitude is

$$F_C = 2F - F = 10^{-5}\text{ N}$$

10. **D**

Let the two parts be q and $(Q - q)$. Then magnitude of force between the two parts is

$$F = k\frac{q(Q-q)}{d^2} = \left(\frac{k}{d^2}\right)(qQ - q^2)$$

For F to be maximum,

$$\frac{dF}{dq} = 0$$

$\left(\frac{k}{d^2}\right)$ being constant, we have

$$\frac{d((qQ - q^2))}{dq} = 0$$

$Q - 2q = 0$

$q = \frac{Q}{2}$

So, the parts are $\frac{Q}{2}$ and $\frac{Q}{2}$.

11. D

When the charge q is at a distance x from Q, the force of repulsion is

$$F = \frac{1}{4\pi\varepsilon_0}\frac{qQ}{x^2}$$

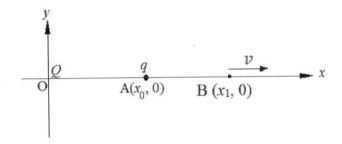

The acceleration is

$$a = \frac{1}{4\pi\varepsilon_0}\frac{qQ}{mx^2}$$

We can write

$$a = \frac{dv}{dt} = \frac{dv}{dx}\frac{dx}{dt} = \frac{dv}{dx}v$$

So,

$$\frac{dv}{dx}v = \frac{1}{4\pi\varepsilon_0}\frac{qQ}{mx^2}$$

$$\int_0^v v\,dv = \frac{1}{4\pi\varepsilon_0}\frac{qQ}{m}\int_{x_0}^{x_1}\frac{1}{x^2}$$

$$\frac{v^2}{2} = \frac{1}{4\pi\varepsilon_0}\frac{qQ}{m}\left[\frac{1}{x_0} - \frac{1}{x_1}\right]$$

$$v^2 = \frac{qQ}{2\pi\varepsilon_0 m}\left[\frac{1}{x_0} - \frac{1}{x_1}\right]$$

12. B

If a charge q were placed at F, by symmetry, the vector sum of forces on charge $-q$ at O due to charges at all the six corners should be zero. Therefore, in absence of charge at F, the vector sum of forces on charge $-q$ due to charge q at A, B, C, D and E should be equal and opposite to the force on charge $-q$ at O due to the q at F.

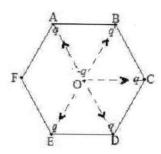

Note that the forces on $-q$ at O due to charges at A and D are equal and opposite (see figure). So, the vector sum of forces on $-q$ at O due to charges at A and D is zero. Similarly, the forces on $-q$ due to charges at D and E, being equal and opposite, cancel out.

Now, the net force on $-q$ at O is due to charge at F and its magnitude is

$$F = k\frac{q^2}{d^2} = \frac{q^2}{4\pi\varepsilon_0 d^2}$$

Being force of attraction, it is directed along OF.

As noted above, the vector sum of forces on charge $-q$ due to charge q at A, B, C, D and E should be equal and opposite to the force on the charge $-q$ at O due to the charge q at F.

Hence, the net force on charge $-q$ at O due to charge q at A, B, C, D and E is directed along FO and its magnitude is

$$F = \frac{q^2}{4\pi\varepsilon_0 d^2}$$

13. C

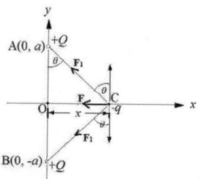

Suppose charge $-q$ is at point C, where OC $= x$.

The magnitudes of the electric force of attraction $\mathbf{F_1}$ on $-q$ due to each of the two charges Q is given as

$$F_1 = k \frac{Qq}{(AC)^2} = k \frac{Qq}{(a^2 + x^2)}$$

Note that the components parallel to AB cancel away whereas the components along CO add up.

The net force $\mathbf{F}$ on $-q$ is along CO and its magnitude is

$$F = 2k \frac{Qq}{(a^2 + x^2)} \cos(90° - \theta) = 2k \frac{Qq}{(a^2 + x^2)} \sin \theta$$

From the figure,

$$\sin \theta = \frac{OC}{AC} = \frac{x}{\sqrt{(a^2 + x^2)}}$$

$$F = 2k \frac{Qq}{(a^2 + x^2)} \frac{x}{\sqrt{(a^2 + x^2)}}$$

$$F = 2k \frac{Qq}{(a^2 + x^2)^{\frac{3}{2}}} x$$

For $x \ll a$,

$$F \approx 2k \frac{Qq}{a^3} x$$

$$F \propto x$$

14. **B**

Suppose the particle with charge $-q$ is to the left of O at point C ($-x$, 0).

In this case, the magnitudes of the electric force of attraction on $-q$ due to each of the two charges Q is given as

$$F_1 = k \frac{Qq}{(a^2 + x^2)}$$

Following arguments similar to those used in previous question, the net force on $-q$ is along CO, i.e. along positive direction of x-axis and its magnitude is

$$F = 2k \frac{Qq}{(a^2 + x^2)^{\frac{3}{2}}} x$$

When the particle is far away ($x \gg a$) to the left of O, F is zero.

Again, when $x = 0$, F is zero.

Thus, F must increase to a maximum value and then decrease to zero at O.

When the particle crosses O, it continues to move towards right. However, the electrostatic force being attractive, the net force on $-q$ is towards O. The direction of net force F is opposite to that of particle's motion and is negative. The magnitude of F remaining the same as above, F is zero both at O and at large distance to the right. Combining the above results, we conclude that the correct choice is (B).

1.2 Electric Field and Electric Potential *and*
1.3 Electric Potential Due to Point Charges and Uniform Fields

15. A

The electric field of magnitude E should be directly upwards so that the electric force on positive charge Q is equal and opposite to the downward gravitational force.

$E Q = m g$

$Q = m \dfrac{g}{E} = \dfrac{(5 \times 10^{-9} \text{ kg}) \times 10 \text{ ms}^{-2}}{10^5 \text{ NC}^{-1}} = 5 \times 10^{-13} \text{ C}$

16. C

The electric field $\mathbf{E}$ at the origin is due to charges Q and $4q$ at A and B respectively. Its magnitude is

$E = k \dfrac{Q}{(a)^2} + k \dfrac{4q}{(2a)^2}$

$E = 0$ gives

$k \dfrac{Q}{(a)^2} + k \dfrac{4q}{(2a)^2} = 0$

$4Q + 4q = 0$

$Q = -q$

17. E

The force on unit charge i.e., electric field is larger at places where electric field lines are nearer to each other than those places where the field lines are farther apart. So, from the diagram, we find that

$F_P > F_Q > F_R$

18. C

The uniform external field $\mathbf{E}$ of magnitude E is directed upward in the plane of the paper.

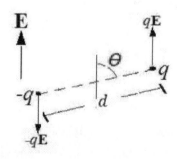

The force on q is $\mathbf{F}_q = q\,\mathbf{E}$

The force on $-q$ is $\mathbf{F}_{-q} = -q\,\mathbf{E}$

So, the net force on the dipole = $\mathbf{F}_q + \mathbf{F}_{-q} = 0$

19. B

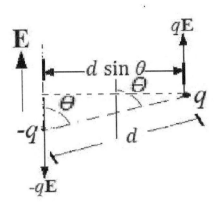

Even though the net force on the dipole is zero, the lines of action of forces on q and $-q$ do not coincide. This fact results in a torque on the dipole.

The magnitude of the torque equals the magnitude of either of the forces multiplied by the perpendicular distance between the two antiparallel forces.

So, the magnitude of torque

$= q E \times d \sin \theta$

$= q d E \sin \theta$

20. A

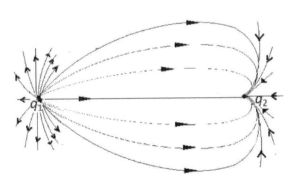

The electric field lines are seen to start from q_1 and end at q_2. Therefore q_1 is positive and q_2 is negative. Since the number of electric field lines per unit area near q_1 is greater than those near q_2, the magnitude of electric field **E** at some distance r near q_1 is stronger than that at the same distance near q_2.

Therefore,

$|q_1| > |q_2|$

21. B

Recall that magnitude of electric field at a distance r from a charge q is $|E| = k\frac{|q|}{r^2}$. So, at a finite distance to the right of q_2, the electric field can be zero because the magnitude of q_2 is smaller than that of q_1 and the distance from q_2 is smaller than that from q_1. Consequently, the electric field created by q_2 at a particular point will cancel out the electric field created by q_1. On the other hand, the electric field intensity cannot be zero to the left of q_1 because the magnitude of q_1 is greater and the distance from q_1 is smaller than that from q_2. Hence, the correct choice is (B).

22. C

Force on a charge q is

$\mathbf{F} = q\,\mathbf{E}$

Work done (W) in moving the charge q along AB

$W = \mathbf{F}\cdot\mathbf{d} = F\,d\cos\theta = q\,E\,d\cos\theta$

$E = \dfrac{W}{q\,d\cos\theta}$

Here, $W = 6$ J, $q = 0.4$ C, $d = 3$ m, $\theta = 60^0$

$E = \dfrac{6\text{ J}}{(0.4\text{ C})(3\text{ m})(\cos 60^0)} = 10.0\,\dfrac{\text{N}}{\text{C}}$

23. C

Potential difference = Work done/Charge

$= \dfrac{\text{Force} \times \text{distance}}{\text{current} \times \text{time}}$

$= \dfrac{(\text{mass} \times \text{acceleration}) \times (\text{distance})}{\text{current} \times \text{time}}$

$= \dfrac{(M\,L\,T^{-2})(L)}{A\,T}$

$= M^1 L^2 T^{-3} A^{-1}$

24. C

Electric field at O is a vector and equals the vector sum of electric field vectors due to q and $-q$.

Electric field vectors at O due to q and $-q$ have the same magnitude ($= k\dfrac{|q|}{a^2}$) and are in the same direction (towards A). They add up at O. So, their vector sum can never be zero.

Electric potential at O is a scalar sum of individual potentials. Electric potential at O due to q and $-q$ have the same magnitude ($= k\dfrac{|q|}{a}$) and opposite signs. So, their scalar sum is zero

25. **B**

Potential at O is

$$V_O = k\frac{q}{a} + k\frac{(-q)}{a} = 0$$

Potential at C is

$$V_C = k\frac{q}{3a} + k\frac{(-q)}{a} = -k\frac{2q}{3a}$$

We know that the work done W by an external force in moving a charge Q from O to C is independent of the path followed and is given in terms of the potential values as

$$W = Q(V_C - V_O)$$

So,

$$W = -k\frac{2Qq}{3}$$

$$= -\frac{Qq}{6\pi\varepsilon_0 a}$$

26. **D**

When charged particle with charge q is far away, its potential energy U_0 is zero and kinetic energy is $K_0 = (1/2)mv^2$. When it moves towards particle with charge Q at A, it experiences force of repulsion. Its kinetic energy decreases and gets converted into potential energy.

Let the distance of the closest approach be AP = d.

At P, kinetic energy K_P of particle with charge q is zero and its potential energy is

$$U_P = k\frac{qQ}{d}$$

Applying conservation of energy,

$$K_0 + U_0 = K_P + U_P$$

$$\left(\frac{1}{2}\right)mv^2 + 0 = 0 + k\frac{qQ}{d}$$

The above equation gives

$$d = k\frac{2qQ}{mv^2} = \frac{qQ}{2\pi\varepsilon_0 mv^2}$$

27. C

In order to find a quick answer to such questions, we can make use of a well-known compact formula, which is as follows.

The potential energy U of a set of n point charges $q_1, q_2, \ldots, q_n$ (which also equals the work done by an external agent to assemble this set of charges) is given as

$$U = k \sum_{i<j} \frac{q_i q_j}{r_{ij}}$$

Here r_{ij} is the fixed distance between q_i and q_j.

For applying this formula to present case, note that

$q_1 = Q, q_2 = q, q_3 = q.$

$r_{12} = AB = a, r_{23} = BC = a$ and $r_{13} = AC = \sqrt{2}\,a.$

Potential energy of the arrangement U

$$= k \left(\frac{q_1 q_2}{r_{12}} + \frac{q_2 q_3}{r_{23}} + \frac{q_1 q_3}{r_{13}} \right)$$

$$= k \left(\frac{Q q}{a} + \frac{q^2}{a} + \frac{Q q}{\sqrt{2}\,a} \right)$$

Since $a \neq 0$, when $U = 0$, we have

$$Q + q + \frac{Q}{\sqrt{2}} = 0$$

$$Q \left(1 + \frac{1}{\sqrt{2}}\right) = -q$$

$$\frac{Q}{q} = -\frac{\sqrt{2}}{\sqrt{2}+1}$$

28. A

Center O is at the same distance $d\ (= a/\sqrt{2})$ from each corner of the square.

The work W, needed by an external force to bring a charge Q to the point O when the four charges are at their respective places is

$W = q_0 \times$ the electrostatic potential at O due to four charges

$$= q_0 \times k \left[\frac{(+q)}{d} + \frac{(-q)}{d} + \frac{(+q)}{d} + \frac{(-q)}{d} \right]$$

The above equation gives

$W = 0$

29. B

The relation

$$E_x = -\frac{dV}{dx}$$

tells us that the electric field is the negative of the slope of the potential V-x curve.

From the given graph, E_x is seen to be positive for $x > 0$ and negative for $x < 0$. So, the potential curve must have non-zero negative slope for for $x > 0$ and positive slope x for < 0.

We find that the graph (B) satisfies this condition.

30. C

The magnitude of electric field is given by the space rate of change of electric potential. In other words,

$$|E| = \frac{|\Delta V|}{\Delta x}$$

The potential difference between the successive lines of constant potential is $\Delta V = 10$ V. However, the perpendicular distance Δx between successive lines is smaller at the point R than at the points P and Q. In other words, the equipotential surfaces are closer at R than at P and Q.

So,

$$\frac{|\Delta V|}{\Delta x} \text{ at R} > \frac{|\Delta V|}{\Delta x} \text{ at P or at Q}$$

Hence, the magnitude of the electric field is greatest at the point R.

31. D

Uniform electric field is in the direction in which the potential decreases.

Therefore,

$V_P > V_Q > V_R$

32. **B**

Electric field being uniform, the magnitude of electric force F on both the charges is

$F = qE$

The acceleration of A and B are respectively

$a_1 = \dfrac{F}{m}$ and $a_2 = \dfrac{F}{M}$

Distance travelled by A in time t_1

$d = \left(\dfrac{1}{2}\right) a\, t_1^2 = \left(\dfrac{1}{2}\right) \dfrac{qE}{m} t_1^2 \qquad (i)$

Similarly, distance travelled by B in time t_2

$d = \left(\dfrac{1}{2}\right) \dfrac{qE}{M} t_2^2 \qquad (ii)$

Dividing eq. (i) by (ii) we get

$1 = \dfrac{M}{m} \dfrac{t_1^2}{t_2^2}$

$\dfrac{t_1}{t_2} = \left(\dfrac{m}{M}\right)^{\frac{1}{2}}$

33. **C**

The electrostatic force is conservative. So, work done by electric field **E** is independent of the path. It depends only on the initial and final points. The work done along the path A to B to C to O is equal to the work done along the direct path from A to O.

Since **E** is uniform, force **F** on the charge is constant.

$\mathbf{F} = q\mathbf{E} = qE\,\mathbf{i}$

We have the position vector of initial point A $(a, 2a, 0)$

$\mathbf{r}_A = a\,\mathbf{i} + 2a\,\mathbf{j}$

Final point is the origin O $(0, 0, 0)$

So, displacement

$\Delta \mathbf{r} = \mathbf{r}_O - \mathbf{r}_A = -a\,\mathbf{i} - 2a\,\mathbf{j}$

Work done by electric field is

$W_E = \mathbf{F}.\Delta \mathbf{r} = (qE\,\mathbf{i}).\Delta \mathbf{r}$

$W_E = (qE\,\mathbf{i}).(-a\,\mathbf{i} - 2a\,\mathbf{j}) = -aqE$

34. B

We shall make use of law of conservation of energy to arrive at the result.

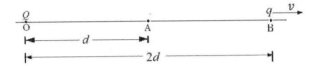

As it moves away from charge Q due to repulsion, potential energy of particle with charge q decreases and gets converted into kinetic energy. The potential energy of the particle at A ($= U_A$) is related to the potential V_A at A as

$U_A = q V_A$

$U_A = q \left(k \dfrac{Q}{d} \right)$

The potential energy at B is

$U_B = q \left(k \dfrac{Q}{2d} \right)$

If v is the speed of particle at B, its kinetic energy is ($\frac{1}{2} m v^2$).

Conservation of energy implies

$\dfrac{1}{2} m v^2 = U_A - U_B = q \left(k \dfrac{Q}{2d} \right)$

$v^2 = \dfrac{2 k q}{m} \left(\dfrac{Q}{2 d} \right)$

$v = \left(\dfrac{q Q}{4 \pi \varepsilon_0 m d} \right)^{\frac{1}{2}}$

35. B

The x-components of electric field is

$E_x = -\dfrac{dV}{dx} = -\dfrac{d(3 x^2)}{dx} = -6x$

Since V is independent of y and z,

$E_y = E_z = 0$

Electric field at P (2, 0, 1)

$\mathbf{E} = -(6x)|_{(2,0,1)} \mathbf{i} = -12 \text{ N C}^{-1} \mathbf{i}$

Charge $q = 2.0$ n C $= 2.0 \times 10^{-9}$ C

So, force on charge q

$\mathbf{F} = q \mathbf{E} = -24.0 \times 10^{-9} \text{ N} \mathbf{i}$

36. A

All points on the circle are at the same distance R from the charge $+q$ at O.

So, electric potential at all points on the circle have a common value

$$V = k\frac{q}{R}$$

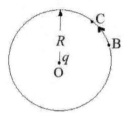

Hence the circle is an equipotential surface.

For points B and C on the circle, $V_C = V_B$

Work done in carrying a charge q_0 from point B to C along the circle

$= q_0(V_C - V_B) = 0$

1.4 Gauss's law

37. A

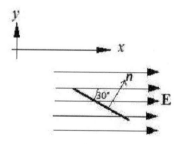

The angle between the plane of the frame and the electric field = 30^0.

So, the angle between the normal n to the plane of the frame and the electric field, $\theta = 90^0 - 30^0 = 60^0$.

Area of the frame

$A = (20 \times 10^{-2} \text{ m})^2 = 4 \times 10^{-2} \text{m}^2$

If **A** is the area vector (normal to the plane of the frame), flux through the frame is

$\varphi = \mathbf{E}.\mathbf{A} = E A \cos\theta$

$= (40.0 \text{ V/m}) \times (4 \times 10^{-2}\text{m}^2) \times \left(\frac{1}{2}\right)$

$\varphi = 0..8 \text{ Vm}$

38. C

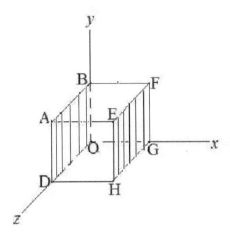

A cube has six faces. As seen in the diagram, four faces are parallel to x-axis.

The electric field, being along x-axis, is parallel to these four faces that are themselves parallel to x-axis. So, for these faces $\mathbf{E}.d\mathbf{A} = E\, dA \cos 90^0 = 0$. Therefore, the flux φ is zero for all of them.

The two faces perpendicular to x direction are ABOD and EFGH.

Between them, for the left face ABCD, the angle between $\mathbf{E}$ and area vector $d\mathbf{A}$ is 180^0 whereas for the right face EFGH, the angle between $\mathbf{E}$ and area vector $d\mathbf{A}$ is 0^0.

Now, $E_x = E_0\, x$

For the left face ABCD, $x = 0$.

So, flux through ABCD is

$\varphi_L = \mathbf{E}.d\mathbf{A} = E_0\,(0) \cos 180^0 = 0$

For the right face EFGH, $x = a$.

So, electric field on the right face is $E = (E_0\, a)$

Flux through the right face is

$\varphi_R = \mathbf{E}.d\mathbf{A} =$

$= E\,(a^2) \cos 0^0 = E\, a^2$

$= (E_0\, a)\, a^2 = E_0\, a^3$

Net flux through the cube

$\varphi = \varphi_L + \varphi_R = 0 + E_0\, a^3$

If q be the charge enclosed in the cubical volume, Gauss's law tells us that

$\varphi = \dfrac{q}{\varepsilon_0}$

$q = \varepsilon_0\, \varphi = \varepsilon_0\, a^3 E_0$

39. B

To solve this problem, we imagine a closed Gaussian surface S as follows.

Place an exactly similar cylindrical vessel upside down on the given vessel as shown below. We thus obtain a closed surface S comprising of two cylinders. The charge enclosed in S is q.

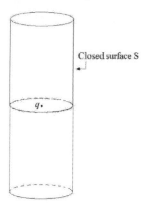

Applying Gauss's law to such a closed surface, we find that the net flux through S is

$$\varphi = \frac{q}{\varepsilon_0}$$

Now, the charge q is situated symmetrically at the center of the two halves of the Gaussian surface S. If φ_c is flux through the given cylindrical vessel, we have

$$\varphi = 2\,\varphi_c$$

So,

$$\varphi_c = \frac{\varphi}{2} = \frac{q}{2\,\varepsilon_0}$$

40. C

Imagine a Gaussian surface in the shape of a cube of edge a with the given square of side a being one of the faces of the cube. Then, as shown below, the given charge q is seen to be placed at the centre of the cube.

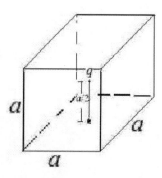

According to Gauss's law, the net electric flux through the whole cube is given by

$$\varphi = \frac{q}{\varepsilon_0}$$

From symmetry, the same value of electric flux emerges through each of six faces. If φ_s is flux through each of the six faces (including the given square), we have

$$\varphi = 6\,\varphi_s$$

$$\varphi_s = \frac{\varphi}{6} = \frac{q}{6\,\varepsilon_0}$$

41. C

Gauss's law tells us that electric flux φ through a closed surface S is

$$\varphi = \oint_S \mathbf{E}\cdot d\mathbf{A} = \frac{q_{enclosed}}{\varepsilon_0}$$

In the above equation, electric flux φ through the closed surface equals $(1/\varepsilon_0)$ multiplied only by the charges enclosed *within* the surface. However, the electric field **E** is the net electric field by *all* the charges present including those outside the surface as well as those enclosed within the surface.

42. B

Net outward electric flux φ through the closed surface = flux leaving the closed surface - flux entering the closed surface.

$$\varphi = \varphi_2 - \varphi_1$$

According to Gauss's law, the net outward electric flux through any closed surface is related to the charge q inside the surface as

$$\varphi = \frac{q}{\varepsilon_0}$$

or, $q = \varepsilon_0\,\varphi = \varepsilon_0(\varphi_2 - \varphi_1)$

43. B

According to Gauss's law, the net electric flux through any closed surface depends only on the charge enclosed by the surface.

$\varphi = (1/\varepsilon_0)(q_{enclosed})$

$\varphi_1 = (1/\varepsilon_0)(3\,q + q - q) = (1/\varepsilon_0)\,(3\,q)$

$\varphi_2 = (1/\varepsilon_0)\,(4\,q + q - 2q) = (1/\varepsilon_0)\,(3\,q)$

$\varphi_3 = (1/\varepsilon_0)\,(2\,q + q) = (1/\varepsilon_0)\,(3\,q)$

$\varphi_4 = (1/\varepsilon_0)\,(2q + 2\,q - q) = (1/\varepsilon_0)\,(3\,q)$

Hence,

$$\varphi_1 = \varphi_2 = \varphi_3 = \varphi_4$$

44. A

Even if the radius is doubled, the charge enclosed will continue to be q. Since, as per Gauss's law, the net electric flux through any closed surface depends only on the charge enclosed by the surface, there will be no effect on the net flux if the radius is doubled.

45. E

To calculate electric field **E** at a point outside the shell, we take the Gaussian spherical surface S with center O and radius r ($> R$).

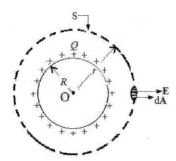

By spherical symmetry, the electric **E** is along the radius vector at each point of S and has the same magnitude E.

Thus, **E** and the area element d**A** at every point are parallel and the flux through each element is $E\, dA \cos 0° = E\, dA$. Summing over all elements of area, the flux through the Gaussian surface is

$$\varphi = \oint \mathbf{E \cdot dA} = E \times (4\pi r^2)$$

If σ is the uniform the charge density, the charge enclosed within S

$q_{enclosed} = \sigma \times (4\pi R^2)$ = the total charge Q on the spherical shell.

By Gauss's law

$$\varphi = \frac{q_{enclosed}}{\varepsilon_0}$$

$$E \times (4\pi r^2) = \frac{Q}{\varepsilon_0}$$

$$E(r) = \frac{Q}{4\pi\varepsilon_0 r^2}$$

The above result continues to be valid right up to the surface of the shell.

To summarise,

$$E(r) = \frac{Q}{4\pi\varepsilon_0 r^2}, r \geq R$$

This expression is exactly the electric field produced by a charge Q placed at the centre O.

To find electric field at a point inside the shell, we again consider a Gaussian spherical surface S´ of radius r and centered at O.

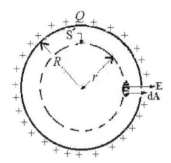

By symmetry arguments similar to those used above, the flux through the Gaussian surface is

$\varphi = \oint E.dA = E \times (4\pi r^2)$

However, in this case, the Gaussian surface encloses no charge.

By Gauss's law

$E(r) \times (4\pi r^2) = \dfrac{q_{enclosed}}{\varepsilon_0} = 0$

$E(r) = 0, r < R$

We find that for points outside the shell, the field due to a uniformly charged shell, being as if the entire charge of the shell is concentrated at its center, varies as $\left(\dfrac{1}{r^2}\right)$.

Inside the shell, the electric field due to a uniformly charged thin shell is zero at all points.

Therefore, graph (E) is the correct choice.

46. C

To calculate electric potential V at a point P outside the shell, we make use of the fact that the charge of the shell can be assumed to be concentrated at the centre of the shell. So, the electric potential at a distance r from the centre of the shell

$V(r) = \dfrac{Q}{4\pi\varepsilon_0 r}, (r > R)$

On the surface of the shell

$V(R) = \dfrac{Q}{4\pi\varepsilon_0 R}, (r = R)$

To find the electric potential V at any point P inside the shell, remember that the electric field **E** is zero at all points inside the shell. Therefore, no work is done in moving a unit test charge from a point on the surface of the shell to the point P inside the shell implying thereby that electric potential at P equals its value at the surface of the shell.

$V(r) = \dfrac{Q}{4\pi\varepsilon_0 R}, (r < R)$

Graph (C) best represents the above results.

47. B

Let the uniform linear charge density be λ. We shall use Gauss's law to calculate the electric field values.

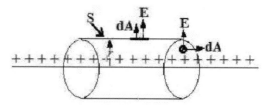

Consider a point P at a radial distance r from the axis of wire. To calculate electric field **E** at P, we take cylindrical Gaussian cylindrical surface S passing through P. By symmetry, the electric field is everywhere radial in the plane cutting the wire normally. Its magnitude depends only on the radial distance r from the wire. The electric flux φ through S can be written as

φ = flux through the two end caps + flux through the curved cylindrical part of the surface

Note that the radial electric field is parallel to the end caps of the closed cylindrical surface S at every point. So, the electric flux through each area element **dA** of end cap is

$E\, dA \cos 90^0 = 0$.

Summing up over all area elements we find that electric flux through whole of end cap = 0.

On the cylindrical part of S, **E** is normal to the surface at every point, and its magnitude is constant. The surface area of the curved part is $2\pi rL$, where L is the length of the cylinder.

Flux through the curved part of S is = $E\,(2\pi rL) \cos 0^0 = E \times 2\pi rL$. Thus,

$\varphi = 0 + E \times 2\pi rL$

Charge enclosed in S = λL

By Gauss's law,

$$\varphi = \frac{q_{enclosed}}{\varepsilon_0}$$

$$E \times 2\pi rL = \frac{\lambda L}{\varepsilon_0}$$

$$E = \frac{\lambda}{2\pi r \varepsilon_0}$$

48. C

To calculate electric field E at a point P distant $r\,(> R)$ outside the uniformly charged sphere, we take the Gaussian spherical surface S with center O and passing through P (see Figure (a) below). By spherical symmetry, the electric E is along the radius vector at each point of S and has the same magnitude E. In other words, **E** is normal to the surface at every point. **E** and the area element d**A** at every point are parallel. So, the flux through each element is $E\, dA \cos 0^0 = E\, dA$. So, the flux through each element is $E\, dA \cos 0^0 = E\, dA$. Summing over all dA, the flux φ through the Gaussian surface is

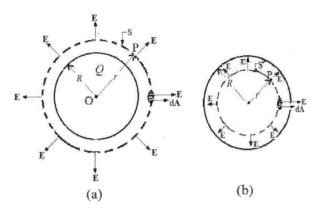

$\varphi = \oint E \cdot dA = E \times (4\pi r^2)$

Since the entire charge Q is contained inside S, Gauss's law gives

$E \times (4\pi r^2) = \dfrac{Q}{\varepsilon_0}$

Because of continuity, the above result remains to be valid on the surface of the shell.

$E = \dfrac{Q}{4\pi \varepsilon_0 r^2}, (r \geq R)$

For point P inside the sphere, we consider a Gaussian spherical surface S' with center O, radius $r (< R)$ and passing through P (Figure (b)). By symmetry arguments similar to those used above, the flux through the Gaussian surface is

$\varphi = \oint E \cdot dA = E \times (4\pi r^2)$

Since charge is distributed uniformly, the charge enclosed within S is

$q_{enclosed} =$ Charge per unit volume $\times$ volume of S

$= \left(\dfrac{Q}{\frac{4}{3}\pi R^3} \right) \left(\dfrac{4}{3}\pi r^3 \right) = \dfrac{Q r^3}{R^3}$

As before, by Gauss's law,

$E \times (4\pi r^2) = \dfrac{Q r^3}{\varepsilon_0 R^3}$

$E = \dfrac{Q r}{4\pi \varepsilon_0 R^3}, (r < R)$

We thus find that for points outside the sphere, the field due to a uniformly charged sphere produces an electric field as if the entire charge of the sphere were concentrated at its center. In other words, for $r > R$, E varies as $\left(\dfrac{1}{r^2}\right)$. Inside the uniformly charged sphere $(r < R)$, the electric field E is directly proportional to r.

Therefore, graph (C) is the only correct choice.

49. A

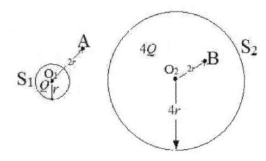

For sphere S_1 of radius r and having charge Q, point A at distance $2r$ from its center is outside the sphere. So,

$$E_1 = \frac{Q}{4\pi\varepsilon_0 (2r)^2} = \frac{Q}{16\pi\varepsilon_0 r^2}$$

For sphere S_2 of radius $4r$ and having charge $4Q$, point, B at distance $2r$ from its center is inside the sphere. Using the result obtained in the previous question, we get

$$E_2 = \frac{(4Q)(2r)}{4\pi\varepsilon_0 (4r)^3} = \frac{Q}{32\pi\varepsilon_0 r^2}$$

We find that $E_1 > E_2$.

50. B

According to Gauss's law, electric flux φ through the closed surface of hollow cylinder equals $(1/\varepsilon_0)$ multiplied only by the charges enclosed within the cylinder.

$$\varphi = \frac{Q}{\varepsilon_0}$$

Now flux through the closed cylindrical surface
= flux through A + flux through B + flux through the curved cylindrical part of the surface

$$\varphi = \varphi_p + \varphi_p + \varphi_c$$
$$= 2 \times \varphi_p + \varphi_c$$
$$\varphi_c = \varphi - 2 \times \varphi_p$$
$$= \frac{Q}{\varepsilon_0} - 2\varphi_p$$

51. D

In this case, the only unique direction is normal to the plane. So, by symmetry, the electric field **E** due to such a sheet must be along this direction. Also, as shown in the side view of the sheet shown below, **E** must have the same magnitude and the opposite directions at two points equidistant from the sheet on opposite sides of it.

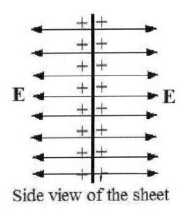

Side view of the sheet

Choose a Gaussian surface S in the form of a closed cylinder with end caps of area A, as shown below.

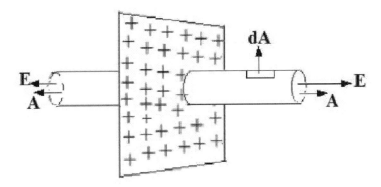

Since electric field **E** is parallel to the cylindrical part of S at every point. As a result, **E** is perpendicular to area elements like **dA** shown in the figure so that **E.dA** = 0. Consequently, there is no flux through the cylindrical portion of the Gaussian surface. The electric field, being normal to the two end caps, the total flux φ through S reduces to the flux through end caps, which becomes

$= E A \cos 0° + E A \cos 0° = 2 E A$.

The charge enclosed by the closed surface is (σ A).

Apply Gauss's law

$$\varphi = \frac{q_{enclosed}}{\varepsilon_0}$$

$$2E A = \frac{\sigma A}{\varepsilon_0}$$

$$E = \frac{\sigma}{2 \varepsilon_0}$$

52. A

The Gaussian surface that we choose for the application of Gauss's law cannot pass through any discrete charge. This is because the electric field is not well defined at the location of discrete charge. It grows towards infinity at distances very close to the discrete charge.

However, the Gaussian surface can pass through a continuous charge distribution.

53. B

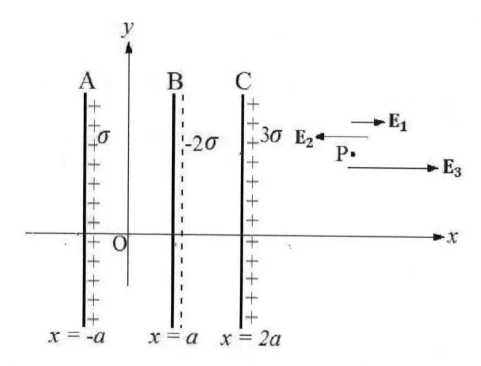

The magnitude of electric field E due to an infinitely large sheet with a uniform surface charge density σ is independent of distance r from the sheet and has the magnitude

$$E = \frac{\sigma}{2\varepsilon_0}$$

Also, the direction of electric field is perpendicular to the sheet. For a positively charged sheets A and C, the electric field values E_1 and E_3 respectively are away from the sheets whereas in case of negatively charged sheet B, the electric field E_2 is towards the sheet.

So, electric field at P is

$$E_P = E_1 + E_2 + E_3$$

$$E_1 = E_1(+i) = \frac{\sigma}{2\varepsilon_0}(+i)$$

$$E_2 = E_2(-\mathbf{i}) = \frac{2\sigma}{2\varepsilon_0}(-\mathbf{i})$$

$$E_3 = E_3(+\mathbf{i}) = \frac{3\sigma}{2\varepsilon_0}(+\mathbf{i})$$

$$E_P = \frac{\sigma}{2\varepsilon_0}\mathbf{i} + \frac{2\sigma}{2\varepsilon_0}(-\mathbf{i}) + \frac{3\sigma}{2\varepsilon_0}\mathbf{i}$$

$$E_P = \frac{2\sigma}{2\varepsilon_0}\mathbf{i} = \frac{\sigma}{\varepsilon_0}\mathbf{i}$$

54. E

The integral ($\oint \mathbf{E} \cdot d\mathbf{A}$) denotes the net electric flux φ through a closed surface.

Now, $\oint \mathbf{E} \cdot d\mathbf{A} = 0$ implies that the net electric flux through a closed surface is zero.

According to Gauss's law, the net electric flux through any closed surface S depends only on the charge enclosed by the surface.

$$\varphi = \oint \mathbf{E} \cdot d\mathbf{A} = \frac{q_{enclosed}}{\varepsilon_0}$$

Note that if **E** is zero at every point of S, ($\oint \mathbf{E} \cdot d\mathbf{A}$) automatically becomes zero. However, even if **E** is not zero at every point of S, the value of integral ($\oint \mathbf{E} \cdot d\mathbf{A}$) can still be zero (for example **E** could be non-zero but parallel to the surface at every point).

If the charge enclosed by S is zero, the flux through S must be zero. So, (E) is the correct choice.

1.5 Fields and potentials of other charge distributions

55. A

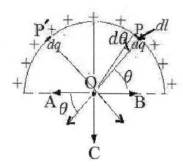

Consider a small charge element of length $dl = (r\, d\theta)$ at point P on the ring. The charge on this element $dq = \lambda\, dl = \lambda\, r\, d\theta$

Electric field at O due to dq is

$$dE = k\,\frac{dq}{r^2} = k\,\frac{\lambda}{r}\,d\theta$$

Note that the components $(dE\cos\theta)$ along OA will be cancelled by another one at point P' of equal magnitude on the opposite side whereas the components along OC add up.

The resultant electric field at O due to semi-circular ring is

$$E = \int dE\, 2\sin\theta\, d\vartheta = 2k\,\frac{\lambda}{r}\int_0^{\frac{\pi}{2}}\sin\theta\, d\vartheta$$

$$= 2k\,\frac{\lambda}{r}[-\cos\theta]_0^{\frac{\pi}{2}} = = 2k\,\frac{\lambda}{r}\left(-(0-1)\right)$$

$$E = 2k\,\frac{\lambda}{r} = \frac{\lambda}{2\pi\varepsilon_0 r}$$

56. B

Consider a small charge element of length dl at point P on the ring.

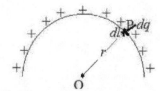

Electric potential dV at O due to dl having charge $dq\ (=\lambda\, dl)$ is

$$dV = k\,\frac{dq}{r} = k\,\frac{\lambda\, dl}{r}$$

Since electric potential is a scalar and all points on the ring are at distance r from O, the resultant potential at O due to semi-circular ring is

$$V = k \frac{\lambda}{r} \int dl = k \frac{\lambda}{r} (\pi r) = k \lambda \pi = \frac{\lambda}{4\varepsilon_0}$$

57. A

To solve this problem, we can make use of the results obtained in the previous questions for semi-circular ring.

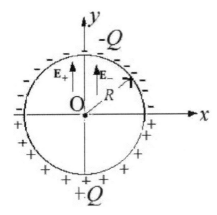

First, note that

$$|\lambda| = \frac{|Q|}{\pi R}$$

The electric field at O due to positively charged semi-circular half part is directed along positive y axis. Its value is

$$\mathbf{E}_+ = \frac{Q}{2\pi^2 \varepsilon_0 R^2} \mathbf{j}$$

The electric field at O due to negatively charged half part is also directed along positive y axis. Its value is again

$$\mathbf{E}_- = \frac{Q}{2\pi^2 \varepsilon_0 R^2} \mathbf{j}$$

So, the net electric field **E** at O due to the whole ring is

$$\mathbf{E} = \mathbf{E}_+ + \mathbf{E}_- = \frac{Q}{\pi^2 \varepsilon_0 R^2} \mathbf{j}$$

Again using the result for electric potential in previous question and noting that electric potential is scalar, we have electric potential V at O

$$V = V_+ + V_- = \frac{Q}{4\pi\varepsilon_0 R} + \frac{(-Q)}{4\pi\varepsilon_0 R} = \frac{1}{4\pi\varepsilon_0 R}(Q - Q) = 0$$

58. D

Charge per unit length $= \dfrac{Q}{L}$

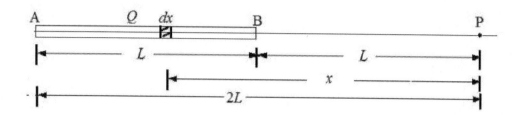

Consider a small charge element dx at a distance x from P. The charge on this element

$$dq = \left(\dfrac{Q}{L}\right) dx$$

Electric potential P due to dq is

$$dV = k \dfrac{dq}{x}$$

To find the potential V_P at P, we have to integrate. Note that for the whole rod, x varies from $x = L$ to $x = 2L$.

$$V_P = \int_L^{2L} k \dfrac{dq}{x}$$

$$V_P = k \left(\dfrac{Q}{L}\right) \int_L^{2L} \dfrac{dx}{x}$$

$$V_P = k \left(\dfrac{Q}{L}\right) [\ln x]_L^{2L}$$

$$= k \left(\dfrac{Q}{L}\right) (\ln 2L - \ln L)$$

$$= k \left(\dfrac{Q}{L}\right) \left(\ln \dfrac{2L}{L}\right)$$

$$V_P = \dfrac{Q}{4 \pi \varepsilon_0 L} (\ln 2)$$

Thus,

$$V_P \propto \dfrac{1}{L}$$

59. B

Consider a small charge element of length dl at point A on the ring as shown. If the charge per unit length is λ, charge on this element

$dq = \lambda \, dl$

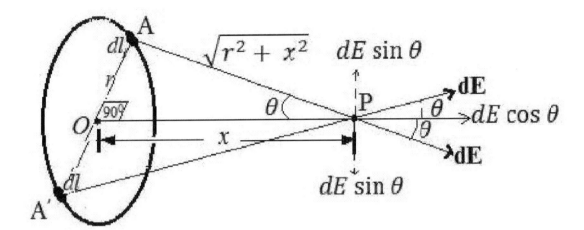

We have

$AP^2 = r^2 + x^2$

Electric field at P due to dq is along AP and has magnitude

$dE = k \dfrac{dq}{(AP)^2}$

$dE = k \dfrac{\lambda \, dl}{(r^2 + x^2)}$

Suppose vector $\mathbf{dE}$ makes an angle θ is with the central axis. It has components $dE \sin\theta$ perpendicular to the axis, which gets cancelled out by an equal and opposite contribution from charge element at diametrically opposite point A'. Such cancellation occurs for all charge elements on the ring.

Thus components of **E** along the axis OP add up and the resultant field has the magnitude

$E = \displaystyle\int dE \cos\theta$

$\cos\theta = \dfrac{x}{\sqrt{r^2 + x^2}}$

$E = k \dfrac{\lambda x}{(r^2 + x^2)^{3/2}} \displaystyle\oint dl$

$E = k \dfrac{\lambda x}{(r^2 + x^2)^{3/2}} (2\pi r)$

Now,

$(2\pi r)\lambda = q$

$E = \dfrac{1}{4\pi\varepsilon_0} \dfrac{qx}{(r^2 + x^2)^{3/2}}$

60. E

Electric potential P due to a small charge element dl at A having charge $dq\ (= \lambda\, dl)$ is

$$dV = k\,\frac{\lambda\, dl}{AP}$$

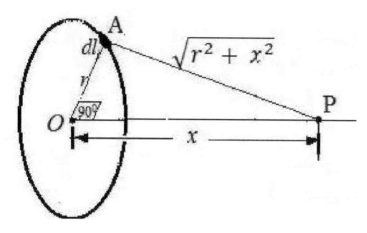

Now, $AP = \sqrt{r^2 + x^2}$

Since electric potential is a scalar and all points on the ring are at distance $\sqrt{r^2 + x^2}$ from P, the resultant potential at P due to the circular ring is

$$V = k\,\frac{\lambda}{\sqrt{r^2 + x^2}}\oint dl = k\,\frac{\lambda}{\sqrt{r^2 + x^2}}\,(2\pi r)$$

Now,

$(2\pi r)\lambda = q$

So,

$$V = \frac{1}{4\pi\varepsilon_0}\,\frac{q}{(r^2 + x^2)^{1/2}}$$

61. B

We use the result obtained above that electric potential V at a point on the axis of a ring of radius r and having charge q at a distance x from its center is

$$V = k\,\frac{q}{(r^2 + x^2)^{1/2}}$$

Potential at O_2 due to charge Q_1 on R_1

$$= k\,\frac{Q_1}{(R^2 + R^2)^{\frac{1}{2}}} = k\,\frac{Q_1}{\sqrt{2}\,R}$$

By putting $x = 0$, potential at O_2 due to charge Q_2 on R_2

$$= k \frac{Q_2}{R}$$

Taking both charges Q_1 and Q_2 into account, we have for the total potential V at O_2

$$V = k \left(\frac{Q_1}{\sqrt{2}\,R} + \frac{Q_2}{R} \right)$$

$$V = \left(\frac{1}{4\pi\varepsilon_0} \right) \left(\frac{Q_1}{\sqrt{2}\,R} + \frac{Q_2}{R} \right)$$

62. E

The integral $\int_{x=\infty}^{x=0} -\mathbf{E}\cdot\mathbf{dx}$ is the work done by an external force in bringing a unit positive charge from infinity to the center of the ring. Such a quantity is, by definition, the electric potential at O.

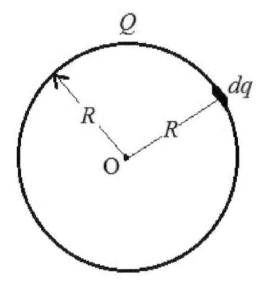

To find the potential at O, let us consider a small charge element dq as shown in the figure. The potential at O due to dq is

$$dV = k \frac{dq}{R}$$

All such small charge elements on the ring are at equal distance R from O. So, even if the charge Q is distributed non-uniformly on the circumference of the ring, the resultant potential at O due to the circular ring is

$$V = k \frac{1}{R} \oint dq = k \frac{1}{R} (Q)$$

$$V = \frac{Q}{4\pi\varepsilon_0 R}$$

UNIT 2: Conductors, Capacitors, Dielectrics

2.1 Electrostatics with Conductors

63. B

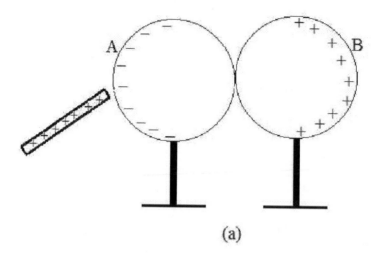

(a)

A glass rod rubbed with silk acquires positive charge. When this positively charged glass rod is brought near A, which is in touch with B, due to process of induction, the free electrons in the spheres are attracted towards the rod. This leaves an excess of positive charge on the surface of sphere B away from the rod. As shown in Figure (a), as long as the sphere are in touch with each other, the surface of sphere A near the rod has an excess of negative charge and the surface of sphere B away from the rod has an excess of positive charge.

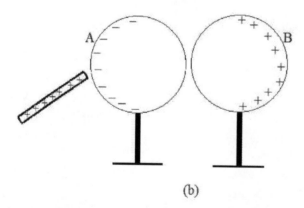

(b)

When A and B are slightly separated (Figure (b)), keeping the glass rod near sphere A, the negative charge on A remains attracted toward the rod and the positive charge on B gets repelled to the surface of sphere B away from the rod.

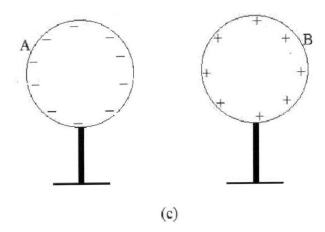

(c)

Finally, when the rod is removed and A and B are separated far apart, the charges on spheres rearrange themselves as shown in Figure (c). The sphere A has negative charge and B has equal amount of positive charge distributed uniformly over them.

64. C

Inside a conductor, electric field is zero. This is because if electric field were not zero, the free electrons in the conductor would experience force and drift causing electric current to flow. So, in the electrostatic situation, electric field must be zero within any conductor.

An immediate consequence of electric field being zero within the conductor is that there cannot be any charge within the conductor. When additional charge Q is given to a conductor, it must reside on the surface of the conductor. This can be seen by applying Gauss's law to any closed surface inside a conductor. Since electric field is zero at every point of such a closed surface, the total electric flux ($= \oint \mathbf{E \cdot dA}$) through it is zero. Hence, Gauss's law tells us that there cannot be any non-zero net charge enclosed by the surface. Consequently, any excess charge must reside at the surface.

Let us see why electric field is normal to S at every point. If it were not normal to the surface, it would have non-zero tangential component along the surface. Consequently, free electrons that experience force would drift and cause electric current to flow along the surface. So, in the static situation, electric field must be normal to S at every point.

Finally, since both electric field within the conductor and tangential component of electric field along the surface are zero, force on any charge is zero. So, no work is done in moving a unit test charge either within the conductor or along the surface of the conductor. Therefore, there cannot be any potential difference between any two points either inside or on the surface of the conductor. Hence, electric potential is constant throughout the volume of the conductor and must have the same constant value as on the surface of the conductor.

It is the Choice (C) that incorporates the above results.

65. C

We know that electric field is zero within the conductor.

$E(r) = 0, r < R$

To find the electric field at a point P at a distance r ($> R$) from O, we note that the charge Q must reside on its outer surface. We consider a Gaussian spherical surface S with centre O and passing through P. From symmetry, electric field **E** should be radial in nature.

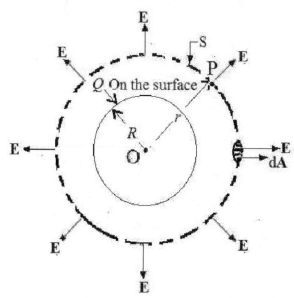

Using spherical symmetry, we find that the flux through S is

$\varphi = \oint \mathbf{E} \cdot \mathbf{dA} = \oint E\, dA\, \cos 0° = E \oint dA = E \times (4\pi r^2)$

Also, the charge Q on surface of the conductor is enclosed within S.
By Gauss's law

$\varphi = \dfrac{q_{enclosed}}{\varepsilon_0}$

$E \times (4\pi r^2) = \dfrac{Q}{\varepsilon_0}$

$E(r) = \dfrac{Q}{4\pi \varepsilon_0 r^2}, r \geq R$

The above results show that (C) is the correct choice.

66. E

Electric potential within any charged metallic conductor has a constant value that equals its value at the surface of the conductor.

To calculate electric potential V at a point P outside the sphere, we make use of the result obtained above that the field outside the sphere is analogous to that of a point charge Q at the center. So the potential outside is that of a point charge at the center of the sphere, i.e.,

$$V(r) = \frac{Q}{4\pi\varepsilon_0 r}, (r \geq R)$$

In particular,

$$V(R) = \frac{Q}{4\pi\varepsilon_0 R}$$

These results show that graph (E) is the correct choice.

67. C

Suppose charge on the shell is Q.

Here, radius of shell, $R = 6$ cm. Point A is within the shell. Since potential within the shell is constant and equals its value at the surface of the conductor. So, potential at the center O = 12 V. Using the formula

$$V(r) = \frac{Q}{4\pi\varepsilon_0 r}, (r \geq R)$$

Potential at the surface

$$12\ V = V(R = 6\text{ cm}) = \frac{Q}{4\pi\varepsilon_0 (6\text{ cm})} \quad (i)$$

Point B is outside the shell. OB = 10 cm

$$V_Q = \frac{Q}{4\pi\varepsilon_0 (10\text{ cm})} \quad (ii)$$

Dividing (ii) by (i),

$$\frac{V_Q}{12\ V} = \frac{6\text{ cm}}{10\text{ cm}}$$

$$V_Q = 7.2\ V.$$

68. C

The charge Q given to conductor resides on the outer surface of the conductor.

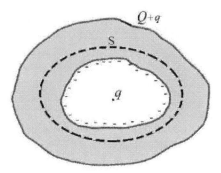

Let us now choose a Gaussian surface S lying wholly within the conductor and enclosing the cavity. Since the electric field at every point inside the conductor is zero, the surface integral of **E** over S (i. e. the flux through S) is

$$\varphi = \oint \mathbf{E}.\,d\mathbf{A} = 0$$

Gauss's law

$$\oint \mathbf{E}.\,d\mathbf{A} = \frac{q_{enclosed}}{\varepsilon_0}$$

tells us that the net charge enclosed in S is zero, which implies that $+q$ within the cavity must induce $-q$ on the inner surface of the conductor. The charge $-q$ on its inner surface must lead to a charge $+q$ on its outer surface. Since the conductor already has a charge Q on its outer surface, the total charge on the outer surface becomes $(Q + q)$.

69. B

First note that (as argued in previous question), the positive charge q in the cavity induces charge $-q$ on the inner surface and a charge $+q$ on its outer surface.

Secondly, the electric field inside the conductor is zero. So, there cannot be any electric field lines inside a conductor. Moreover, the surface of a charged conductor being equipotential, electrostatic field lines must be normal to its surface at every point. So, diagram (B) is the correct choice.

70. B

The electric field inside the conductor is zero and, therefore, there cannot be any electric field lines within the conductor. At the surface of a conductor, electric field lines must be normal to the surface at every point. So, path 2 and hence (B) is the correct choice.

71. **D**

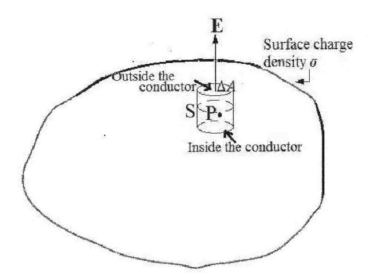

We know that the surface of a conductor is equipotential surface and hence the electric field is perpendicular to the surface. Choose a Gaussian surface S around P in the form of a small closed cylinder with end caps of small area of cross-section ΔA. As shown in the figure, S is partly inside and partly outside the surface of the conductor.

Just outside the surface of the conductor having positive charge, electric field is away from the surface. Inside the surface, **E** is zero. Since field lines are parallel to the cylindrical part of S at every point, there is no flux through this portion of S. So, the contribution to the total flux through the Gaussian surface S comes only from the end cap outside the surface and its value is $\mathbf{E}.\Delta\mathbf{A} = E\Delta A$.

The charge enclosed by the closed surface S is $\sigma \Delta A$.

Apply Gauss's law

$$\varphi = \frac{q_{enclosed}}{\varepsilon_0}$$

$$E\,\Delta A = \frac{\sigma\,\Delta A}{\varepsilon_0}$$

$$E = \frac{\sigma}{\varepsilon_0}$$

72. D

We know that any extra charge given the uncharged conductor will reside on the surface of the conductor-the interior of any conductor must have zero net charge. This eliminates Choice (C). Also, electric field will be normal to the surface of the conductor at every point of the surface. Therefore, Choice (B) cannot be correct. To choose among the remaining three, consider the following.

If a charge Q charge is given to a conducting sphere of radius r, the potential on its surface has a constant value given by

$$V = \frac{Q}{4\pi\varepsilon_0 r}$$

Charge density σ is given as

$$\sigma = \frac{Q}{4\pi\varepsilon_0 r^2}$$

In terms of charge density, we have

$$V = \sigma r$$

Recall that the surface of any charged conductor-including that of the conducting sphere-is equipotential. In other words,

$$V = \sigma r = \text{constant}$$

$$\sigma \propto \frac{1}{r}$$

Charge density is inversely proportional to the radius of the sphere.

Now, consider the given non-spherical charged conductor. As shown below, a portion of its surface, where the surface is more curved, can be considered to be part of a sphere of smaller radius whereas another flatter part can be considered to be part of a sphere of larger radius.

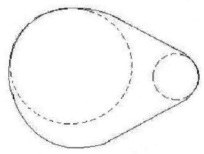

Even though its surface as a whole is equipotential, the charge density σ will not be uniform. In view of above relation, σ will be larger in the part that is more curved and will be smaller in the flatter part.

Finally, the magnitude of electric field near the part of the surface of the conductor where the surface density is σ is

$$E = \frac{\sigma}{\varepsilon_0}$$

In the present case, since σ is larger where surface is more curved than where the surface is 'flat', electric field will be stronger near the part of smaller radius of curvature than near the part of larger radius of curvature. Hence, the correct answer is Choice (D).

73. **B**

We know that the charge given to any conductor lies on its surface. For the thin conducting sheet, both surfaces have uniform charge density σ.

The electric field near a uniformly charged conducting sheet S is normal to the surface and has a magnitude E given by

$$E = \frac{\sigma}{\varepsilon_0}$$

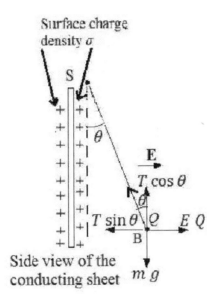

The forces acting on ball B are

(i) the magnitude of electric force $F_e = E\,Q$ along normal to S

(ii) the magnitude of force of gravity mg acting downward

(ii) the tension T along the thread

When equilibrium is reached, the thread makes an angle θ with S. Then from free body diagram of the ball,

$$T \sin \theta = F_e = \frac{\sigma}{\varepsilon_0} Q \quad \ldots (i)$$

$$T \cos \theta = m g \quad \ldots (ii)$$

Division of Eq. (i) by Eq. (ii) gives

$$\tan \theta = \frac{\sigma Q}{\varepsilon_0 mg}$$

74. D

Let E be the (magnitude of) electric field.

For $r < a$, charge enclosed is zero. So, $E(r) = 0$

This result rules out (A) and (B).

Charge on the surface of inner shell is q. So, for $r = a$,

$$E(a) = \frac{q}{4\pi\varepsilon_0 a^2}$$

For $a < r < b$, charge enclosed is q.

$$E(r) = \frac{q}{4\pi\varepsilon_0 r^2}$$

For $r \geq b$, charge enclosed is $|q| + (-|Q|)$. So,

$$E(b) = \frac{\{|q| + (-|Q|)\}}{4\pi\varepsilon_0 b^2}$$

$$E(b) = -\frac{(|Q| - |q|)}{4\pi\varepsilon_0 b^2}$$

The above result shows that for $r = b$, E is negative.

For $r > b$, E continues to be negative and its value is

$$E = -\frac{(|Q| - |q|)}{4\pi\varepsilon_0 r^2}$$

These results are best represented by graph in (D).

2.2 Capacitors

75. D

In the given network, C_1 and C_2 are connected in series. The effective capacitance C_s of these two capacitors is given by

$$\frac{1}{C_s} = \frac{1}{C_1} + \frac{1}{C_2} = \frac{C_1 + C_2}{C_1 C_2}$$

$$C_s = \frac{C_1 C_2}{C_1 + C_2} = \frac{(C)(2C)}{(C + 2C)} = \frac{2C}{3}$$

Charge on C_s is

$$Q_s = C_s V = \frac{2CV}{3}$$

The charge on each of the capacitors C_1 and C_2 in series is also Q_s.

The potential difference across C_3 is V.

The charge on C_3 is

$$Q_3 = C_3 V = (3C) V$$

The ratio of the charges on C_1 and C_3 is

$$\frac{Q_1}{Q_3} = \frac{Q_s}{Q_3} = \frac{2}{9}$$

76. **A**

We number the plates from 1 to 4. Let the points P and Q to be joined to positive and negative terminals respectively of a battery. The plates will acquire charges as shown below.

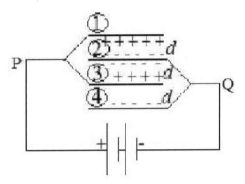

Careful observation of the above figure tells us that the arrangement of plates is equivalent to three parallel plate capacitors each of capacitance (say) C joined in parallel to each other.

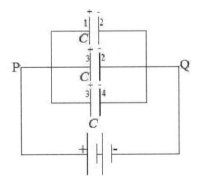

With given parameters,

$$C = \varepsilon_0 \frac{A}{d}$$

So, for parallel combination, capacitance between P and Q is

$$C_{PQ} = C + C + C$$

$$C_{PQ} = 3C = 3\varepsilon_0 \frac{A}{d}$$

77. C

Suppose capacitance of each of the three capacitors is C'. Let them be joined in series between A and B.

The equivalent capacitance of the series combination is C. Therefore,

$$\frac{1}{C} = \frac{1}{C'} + \frac{1}{C'} + \frac{1}{C'} = \frac{3}{C'}$$

$C' = 3C$

So, when they are joined in parallel, the equivalent capacitance is

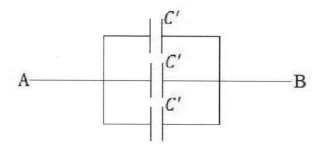

$C_p = C' + C' + C' = 9C$

78. B

The surface charge density of A is $\sigma_A = \dfrac{Q}{A}$ and that of B is $\sigma_B = \dfrac{q}{A}$.

The electric field, in the region between the plates, due to the A has the magnitude

$$E_A = \frac{\sigma_A}{2\varepsilon_0} = \frac{Q}{2\varepsilon_0 A}, E_A \text{ is directed from A to B.}$$

and the electric field due to the B has the magnitude

$$E_B = \frac{\sigma_B}{2\varepsilon_0} = \frac{q}{2\varepsilon_0 A}, E_A \text{ is directed from B to A.}$$

The net electric field in the region between the plates is vector sum of $\mathbf{E}_A$ and $\mathbf{E}_B$

$$\mathbf{E} = \mathbf{E}_A + \mathbf{E}_B$$

Since $Q > q$, $\mathbf{E}$ is directed from A to B and its magnitude is

$$E = \frac{(Q-q)}{2\varepsilon_0 A}$$

The potential difference V between A and B equals the electric field multiplied by the separation between the plates.

$$V = Ex = \frac{(Q-q)}{2\varepsilon_0 A} x = \frac{(Q-q)}{2\left(\varepsilon_0 \frac{A}{x}\right)}$$

Note that the capacitance of parallel plate capacitor having two plates each of area A and separation x is independent of the charge on the plates and depends only on geometrical configuration (area, separation) of the two plates. Its value is

$$C = \varepsilon_0 \frac{A}{x}$$

So, we have

$$V = \frac{(Q-q)}{2C}$$

79. B

Let A be area of each plate and d be the distance between the plates.

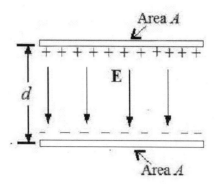

We know that if C be the capacitance and V be the potential difference, the energy stored in the capacitor is

$$U = \frac{1}{2} C V^2$$

For parallel plate capacitor,

$$C = \varepsilon_0 \frac{A}{d}, \quad V = E\,d$$

$$U = \frac{1}{2} \left(\varepsilon_0 \frac{A}{d} \right) (E\,d)^2 = \frac{1}{2} (\varepsilon_0 E^2)(A\,d)$$

Now, $(A\,d)$ is the volume of the region between the plates where the electric field exists.

So, energy density of electric field

$$= \frac{\text{energy stored}}{\text{volume}} = \frac{U}{(A\,d)} = \frac{1}{2} (\varepsilon_0 E^2)$$

80. B

Let A be area of each plate of the parallel plate capacitor and d be the distance between the plates.

Since the charging battery is disconnected after charging, charge Q on the plates remains constant.

Now, for parallel plate capacitor,

$$C = \varepsilon_0 \frac{A}{d}$$

When d increases, capacitance C decreases

Since $V = \frac{Q}{C}$, decrease of C with constant Q implies that V increses.

Energy

$$U = \frac{1}{2} \frac{Q^2}{C}$$

Again, decrease of C with constant Q implies that U increses.

81. E

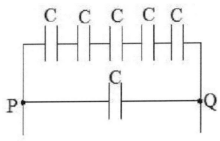

To find the capacitance between P and Q, note that one capacitor is parallel to a series combination of five capacitors.

The equivalent capacitance of five capacitors in series C' is given as

$$\frac{1}{C'} = \frac{1}{C} + \frac{1}{C} + \frac{1}{C} + \ldots \text{ five times } = \frac{5}{C}$$

$$C' = \frac{C}{5}$$

So, capacitance between P and Q is

$$C_{PQ} = \frac{C}{5} + C = \frac{6}{5}C$$

Again, to find the capacitance between P and R, note that a series combination of two capacitors in series is parallel to a series combination four capacitors.

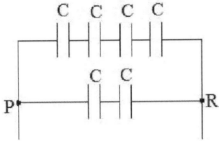

Quick calculation similar to above shows that the equivalent capacitance of two capacitors in series is $\left(\frac{C}{2}\right)$, whereas the equivalent capacitance of four capacitors in series is $\left(\frac{C}{4}\right)$.

Capacitance between P and R is

$$C_{PR} = \frac{C}{2} + \frac{C}{4} = \frac{3}{4}C$$

Therefore, required ratio is

$$\frac{C_{PQ}}{C_{PR}} = \frac{\left(\frac{6}{5}C\right)}{\left(\frac{3}{4}C\right)} = \frac{8}{5}$$

82. D

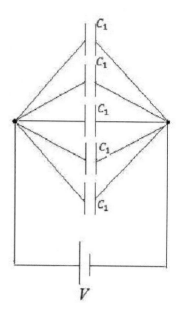

The equivalent capacitance of parallel combination 5 capacitors is

$C_p = C_1 + C_1 + \cdots 5 \text{ times} = 5\, C_1$

The total stored energy of parallel combination

$U_p = \dfrac{1}{2} C_p V^2 = \dfrac{1}{2}(5\, C_1) V^2 = \dfrac{5}{2} C_1 V^2$

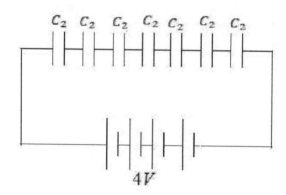

The equivalent capacitance of series combination 7 capacitors is C_s where

$\dfrac{1}{C_s} = \dfrac{1}{C_2} + \dfrac{1}{C_2} + \cdots 7 \text{ times} = \dfrac{7}{C_2}$

$C_s = \dfrac{C_2}{7}$

The series combination is charged to potential difference of $4V$. So, the total stored energy of series combination

$$U_s = \frac{1}{2} C_s (4V)^2$$

$$= \frac{1}{2} \left(\frac{C_2}{7}\right)(4V)^2 = \frac{8}{7} C_2 V^2$$

$U_p = U_s$ gives

$$\frac{5}{2} C_1 V^2 = \frac{8}{7} C_2 V^2$$

So,

$$\frac{C_1}{C_2} = \frac{\frac{8}{7}}{\frac{5}{2}} = \frac{16}{35}$$

83. **B**

As in previous answers, the equivalent capacitance of three identical capacitors of capacitance C (= $6\,\mu F$) in series C_S is $(C/3)$. In this case

$$C_S = \frac{6\,\mu F}{3} = 2\,\mu F.$$

So, (A) is incorrect.

For parallel combination, capacitances add up. The equivalent capacitance is

$$C_p = 3C = 18\,\mu F$$

This shows that (C) is also incorrect.

In choice (D), one of the three capacitors of $6\,\mu F$ is one of the capacitors in a series combination with the parallel combination of other two. We expect the resultant to be less than $6\,\mu F$. The actual calculation gives the resultant as

$$\frac{1}{C_s} = \frac{1}{6\,\mu F} + \frac{1}{(6\,\mu F + 6\,\mu F)} = \frac{1}{6\,\mu F} + \frac{1}{12\,\mu F}.$$

$C_s = 4\,\mu F$ as per our expectation. Therefore, (D) is incorrect.

Careful thought tells us that in order to get a capacitance of $9\,\mu F$, one of the three capacitors of $6\,\mu F$ can be in a parallel combination. The remaining $3\,\mu F$ (= $9\,\mu F - 6\,\mu F$) is obtainable by joining the other two given capacitors in series, which is the case in (B). Therefore, (B) is the correct choice.

84. A

When capacitor C is charged to potential V, initial charge on the capacitor is
$Q_i = CV$
and its initial electrostatic energy is
$$U_i = \frac{1}{2}\frac{Q_i^2}{C}$$

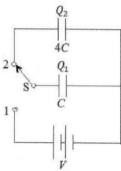

When S is turned to position 2, the two capacitors C and $4C$ are connected in parallel. Let Q_1 and Q_2 be the charges on C and $4C$ respectively. Since two capacitors in parallel have a common potential V',

$$V' = \frac{Q_1}{C} = \frac{Q_2}{4C} = \frac{Q_1 + Q_2}{C + 4C} \quad \ldots\ldots (i)$$

Conservation of charge gives
$$Q_i = Q_1 + Q_2 \quad \ldots\ldots\ldots (ii)$$

From Eqs. (i) and (ii),
$$Q_1 = \frac{Q_i}{5} \text{ and } Q_2 = \frac{4Q_i}{5}$$

Final energies stored in the capacitors are
$$U_1 = \frac{1}{2}\frac{Q_1^2}{C} = \frac{Q_i^2}{50C} = \frac{1}{25}\left(\frac{1}{2}\frac{Q_i^2}{C}\right) = \frac{U_i}{25}$$

$$U_2 = \frac{1}{2}\frac{Q_2^2}{(4C)} = \frac{16Q_i^2}{200C} = \frac{4}{25}\left(\frac{1}{2}\frac{Q_i^2}{C}\right) = \frac{4U_i}{25}$$

Total final electrostatic energy
$$U_f = U_1 + U_2 = \frac{U_i}{5}$$

We thus find that the electrostatic energy decreases. The decrease is
$$= U_i - U_f = \frac{4U_i}{5}$$

Percentage decrease in electrostatic energy = (Decrease in electrostatic energy/ initial energy) × 100

$$= \frac{\left(\frac{4U_i}{5}\right)}{U_i} \times 100 = 80\%$$

85. **B**

When charge $+Q$ is given to the inner cylinder, a charge $-Q$ is induced on the inner surface of the outer cylinder. Since $b \ll L$, cylinder can be considered to be extremely large.

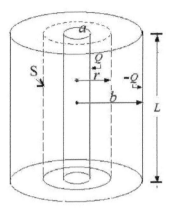

From symmetry, the electric field near the inner cylinder is everywhere radial in the plane cutting the cylinder normally. Its magnitude E depends only on the radial distance r from the wire.

For calculating E by using Gauss's law, we take the Gaussian surface to be a cylinder S of radius r and length L as shown above. **E** being radial (and hence parallel to end caps), flux through the two end caps is zero and through cylindrical part of the surface is $\varphi = E \times 2\pi rL$.

Gauss's law gives

$$\varphi = \frac{q_{enclosed}}{\varepsilon_0}$$

$$E \times 2\pi rL = \frac{Q}{\varepsilon_0}$$

$$E = \frac{Q}{2\pi \varepsilon_0 L\, r}$$

The inner cylinder, being positively charged, is at a higher potential than the outer cylinder. Potential difference V between the cylinders is

$$V = V_a - V_b = -\int_b^a E\, dr = -\frac{Q}{2\pi \varepsilon_0 L}\int_b^a \frac{dr}{r}$$

$$= -\frac{Q}{2\pi \varepsilon_0 L}[\ln r]_b^a = -\frac{Q}{2\pi \varepsilon_0 L}(\ln a - \ln b) = = \frac{Q}{2\pi \varepsilon_0 L}\ln\frac{b}{a}$$

By definition, capacitance is

$$C = \frac{Q}{V} = \frac{2\pi \varepsilon_0 L}{\ln\frac{b}{a}}$$

2.3 Dielectrics

86. B

When a slab of dielectric is introduced, the electric field $\mathbf{E}_0$ that existed between the plates before the introduction of the slab causes polarization of the dielectric. As a result, a layer of negative charge is formed near the positive plate and a layer of positive charge is formed near the negative plate of the capacitor. The field produced by these surface charges opposes $\mathbf{E}_0$. The net field in the dielectric is, thereby, reduced from its earlier value. The reduction factor equals the dielectric constant K of the material of the dielectric. If we denote the magnitude of net electric field in the dielectric as E_K, we have

$$E_K = \frac{E_0}{K}$$

$$E_K \propto \frac{1}{K}$$

Since $K_1 > K_2$, we find that the electric field in S_1 is less than that in S_2.

Hence, graph in (B) correctly depicts the variation of electric field with x.

87. C

When there is no dielectric, the electric field E_0 between the plates is related to the potential difference V_0 between the plates as

$$V_0 = E_0\, d$$

The charge on the plates is related to the capacitance C_0 as

$$Q_0 = C_0 V_0$$

$$C_0 = \frac{Q_0}{V_0} \quad \ldots (i)$$

When the dielectric of thickness t is inserted, the electric field in the dielectric will change. It will have the reduced magnitude

$$E = \frac{E_0}{K}$$

The electric field in the remaining space of thickness $(d-t)$ will continue to be E_0.

The potential difference between the plates will have the changed value

$$V = E\,t + E_0\,(d-t) = \frac{E_0}{K}t + E_0\,(d-t)$$

$$V = E_0 \left(\frac{t}{K} + (d-t)\right) = \frac{E_0\, d}{K}\left(\frac{t}{d} + K\left(1 - \frac{t}{d}\right)\right)$$

So,

$$V = \frac{V_0}{K}\left(\frac{t}{d} + K\left(1 - \frac{t}{d}\right)\right)$$

Since the charge on the capacitor continues to be Q_0 even after introduction of dielectric, we have

$$C = \frac{Q_0}{V} = \frac{K}{\left(\frac{t}{d} + K(1 - \frac{t}{d})\right)} \frac{Q_0}{V_0}$$

Using Eq. (i), we get

$$C = \frac{K}{\left(\frac{t}{d} + K(1 - \frac{t}{d})\right)} C_0$$

$$\frac{C_0}{C} = \frac{t}{dK} + 1 - \frac{t}{d} = 1 - \frac{t}{dK}(K - 1)$$

88. **E**

As above, when there is no metal plate, the potential difference is

$$V_0 = E_0 d$$

and the capacitance without the metal plate is

$$C_0 = \frac{Q_0}{V_0}$$

Recall that the electric field within a conductor is zero. Therefore, the potential difference across width t of the conductor is zero. Since electric field in the remaining space of thickness $(d - t)$ is E_0, the potential difference between the plates is

$$V = \text{zero} + E_0(d - t) = E_0(d - t)$$

$$V = E_0 d \left(1 - \frac{t}{d}\right) = V_0 \left(1 - \frac{t}{d}\right)$$

So, in this case

$$C = \frac{Q_0}{V} = \frac{Q_0}{V_0\left(1 - \frac{t}{d}\right)} = \frac{C_0}{\left(1 - \frac{t}{d}\right)}$$

89. D

Capacitance of parallel plate capacitor with vacuum being the intervening medium between the plates is

$C_0 = \varepsilon_0 \dfrac{A}{d}$

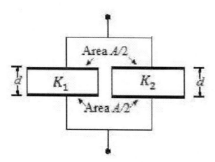

Careful observation shows that the arrangement is equivalent to two capacitors joined in parallel. Each of the two capacitors has plate area $(A/2)$ and separation d. The dielectric constants of the dielectric between the plates of the respective capacitors are K_1 and K_2. If C_1 and C_2 are the capacitance values, we have

$C_1 = \varepsilon_0 K_1 \dfrac{(A/2)}{d}$

$C_2 = \varepsilon_0 K_2 \dfrac{(A/2)}{d}$

So, the capacitance of the arrangement

= equivalent capacitance of parallel combination of C_1 and C_2

$C = C_1 + C_2$

$C = \dfrac{1}{2}(K_1 + K_2)\left(\varepsilon_0 \dfrac{A}{d}\right)$

$C = \dfrac{1}{2}(K_1 + K_2)C_0$

$\dfrac{C}{C_0} = \dfrac{1}{2}(K_1 + K_2)$

90. A

We know that when a dielectric slab having dielectric constants K is introduced to fill the space between the plates, the capacitance of the capacitor becomes C_K, which is related to C_0 as

$C_K = K C_0$

Since $K > 1$, we have

$C_K > C_0$

Because the battery is not removed, the potential across the capacitor continues to be V_0 even after the introduction of the dielectric. So,

$V_K = V_0$

Now, the charge on the capacitor plates is

$Q_K = C_K V_K = (K C_0) V_0 = K (C_0 V_0) = K Q_0$

So,

$Q_K > Q_0$

Electric field after the introduction of the dielectric is

$E_K = \dfrac{V_K}{d} = \dfrac{V_0}{d} = E_0$

Finally,

$U_K = \dfrac{1}{2} C_K V_K^2 = \dfrac{1}{2} (K C_0) V_0^2 = K(\dfrac{1}{2} C_0 V_0^2)$

$U_K = K U_0$

Hence,

$U_K > U_0$

Combining all the above results, we find that correct choice is (A).

B. FREE-RESPONSE QUESTIONS

91. (a)

(i) When equilibrium is reached, let the separation between the balls be x.

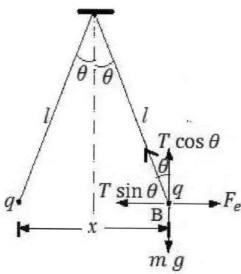

Electrostatic force between the balls

$$F_e = k \frac{q^2}{x^2}$$

Then from free body diagram of either ball,,

$T \sin \theta = F_e$ (i)

$T \cos \theta = mg$ (ii)

Division gives

$$\tan \theta = \frac{F_e}{mg}$$

From the figure,

$$\sin \theta = \frac{x/2}{l}$$

Using $\tan \theta \approx \sin \theta$ and substituting the values, we get

$$\frac{x}{2l} = \frac{F_e}{mg} = k \frac{q^2}{mgx^2}$$

$$x^3 = k \frac{2l q^2}{mg}$$

$$x = \left(k \frac{2l q^2}{mg} \right)^{1/3} = \left(\frac{l q^2}{2 \pi \varepsilon_0 mg} \right)^{1/3}$$

(ii) To obtain tension in the string, we square and add Eqs. (i) and (ii) to get

$$T^2(\sin^2\theta + \cos^2\theta) = \left(k\frac{q^2}{x^2}\right)^2 + m^2g^2$$

$$T = \left[\left(\frac{q^2}{4\pi\varepsilon_0 x^2}\right)^2 + m^2g^2\right]^{1/2}$$

(b)

From above answer,

$$x^3 = \frac{l\,q^2}{2\pi\varepsilon_0\,mg} \quad (i)$$

Differentiating with respect to t, we get

$$3x^2\frac{dx}{dt} = \left(\frac{l}{2\pi\varepsilon_0\,mg}\right)\left(2q\frac{dq}{dt}\right)$$

Noting that $dx/dt = v$, we have

$$x^2 v \propto 2q\frac{dq}{dt} \quad (ii)$$

But rate of leakage of charge i.e., dq/dt is constant
From eq. (i),

$$q \propto x^{3/2}$$

So, from eq. (ii)

$$v \propto q\,x^{-2} = x^{3/2} x^{-2}$$

Hence

$$v \propto \frac{1}{\sqrt{x}}$$

(c)

In absence of gravitational force, there is only electrostatic force.

The electrostatic force of repulsion will push the two balls as far as possible.

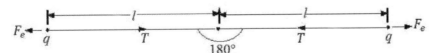

So, the angle θ between the two strings will be 180° and the separation between the balls $x = 2l$

From the free body diagram, we find that the tension in the string will be equal to the electrostatic force of repulsion.

$$T = F_e = k\frac{q^2}{x^2} = \frac{q^2}{16\pi\varepsilon_0 l^2}$$

92. (a)

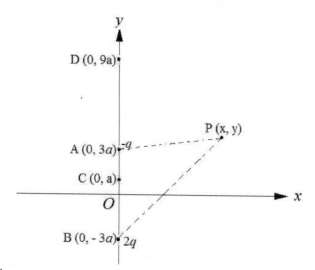

We have the distances

$PA = (x^2 + (y - 3a)^2)^{1/2}$

$PB = (x^2 + (y + 3a)^2)^{1/2}$

Electric potential at P (x, y) due to charges at A and B is

$V(x, y) = k \left(\dfrac{-q}{PA} + \dfrac{2q}{PB} \right)$

$V(x, y) = k q \left(\dfrac{-1}{(x^2 + (y - 3a)^2)^{1/2}} + \dfrac{2}{(x^2 + (y + 3a)^2)^{1/2}} \right)$

(b)

$V(x, y) = 0$, when

$\left(\dfrac{-1}{(x^2 + (y - 3a)^2)^{1/2}} + \dfrac{2}{(x^2 + (y + 3a)^2)^{1/2}} \right) = 0$

$2 (x^2 + (y - 3a)^2)^{1/2} = (x^2 + (y + 3a)^2)^{1/2}$

$4 (x^2 + (y - 3a)^2) = (x^2 + (y + 3a)^2)$

On simplification, we get

$3 x^2 + 3 y^2 - 30 a y + 27 a^2 = 0$

So,

$x^2 + y^2 - 10 a y + 9 a^2 = 0$ (i)

The above is the locus of all those points where electric potential is zero.

(c)

(i) To find the coordinates of all points having zero electric potential on the x-axis, put $y = 0$ in Eq. (i) above,

$x^2 + 9a^2 = 0$

There is no real solution of above equation, which means that there is no point on x-axis where electric potential is zero.

(ii) To find such points on the y- axis, put $x = 0$

$y^2 - 10ay + 9a^2 = 0$

$(y - a)(y - 9a) = 0$

y = a or y = 9a

Thus C (0, a) and D (0, 9a) are two points on the y- axis, where electric potential is zero.

(d)

C $(0, a)$ is the point nearest to origin O, where electric potential is zero.

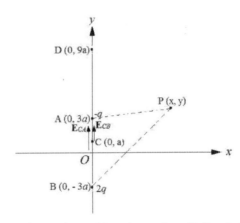

The electric field vector $\mathbf{E}_{CB}$ at C due to the positive charge $2q$ at B (0, - 3a) points towards the positive direction of y-axis and has a magnitude

$$E_{CB} = k\frac{2q}{(3a + a)^2} = k\frac{q}{8a^2}$$

The electric field vector $\mathbf{E}_{CA}$ at C due to the negative charge -q at A (0, 3a) also points towards the positive direction of y-axis and has a magnitude

$$E_{CA} = k\frac{q}{(3a - a)^2} = k\frac{q}{4a^2}$$

So, the net electric field **E** at C is

$\mathbf{E} = \mathbf{E}_{CB} + \mathbf{E}_{CA}$

$\mathbf{E} = k\dfrac{q}{8a^2}\mathbf{j} + k\dfrac{q}{4a^2}\mathbf{j} \Rightarrow \mathbf{E} = \dfrac{3kq}{8a^2}\mathbf{j}$

93. (a)

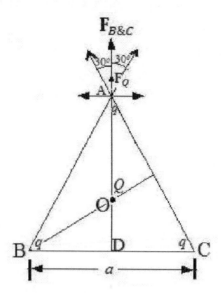

The forces on charge q at A due to charges at B and C are along BA and CA respectively as shown. Each of these forces has magnitude

$$F = k \frac{q^2}{a^2}$$

Clearly, the components parallel to BC axis cancel away. The components along perpendicular to BC add up. So, the resultant of forces on charge q at A due to charges at B and C is

$$F_{B\&C} = 2 F \cos 30^0 = \sqrt{3} k \frac{q^2}{a^2}$$

The force on q at A due to charge Q at O is

$$F_Q = k \frac{qQ}{(OA)^2}$$

Let AD be the perpendicular from A to BC. From geometry, OA = (2/3) AD = (2/3) ($a \cos 30^0$) = (1/√3) a.

$$F_Q = k \frac{qQ}{(\frac{1}{\sqrt{3}}a)^2} = 3k \frac{qQ}{a^2}$$

Net force on q at A is

F = **F**$_{B\&C}$ + **F**$_Q$

When Q and q are of the same sign, **F** is directed along OA and its magnitude is

$$F = k \frac{q}{a^2} (\sqrt{3}\, q + 3\, Q)$$

(b)

For the charge q at A to remain stationary,

$F = 0$, which gives

$3Q = -\sqrt{3}\, q$

$\dfrac{Q}{q} = -\dfrac{1}{\sqrt{3}}$

The above result implies Q and q are of opposite sign.

For an equilateral triangle, OA = OB = OC. Also, charge at the three vertices is equal ($= q$).

Therefore, for $Q/q = -(1/\sqrt{3})$, the charges at A, B and C will remain stationary.

(c)

$Q/q = -1$ gives $Q = -q$.

Putting $Q = -q$ in the expression for F in part (a) above, we get

$$F = k\dfrac{q^2}{a^2}(\sqrt{3} - 3)$$

The above result shows that F is negative, which means that net force $\mathbf{F}$ on A is directed towards O.

This result implies that for $Q/q = -1$, the charge at A and (by symmetry) at B and at C will all move towards the centroid O.

(d)

We use the formula for the potential energy U of a set of n point charges $q_1, q_2, \ldots, q_n$ as

$$U = k \sum_{i<j} \dfrac{q_i\, q_j}{r_{ij}}$$

where r_{ij} is the distance between q_i and q_j.

Here, $q_1 = q_2 = q_3 = q$, $q_4 = Q$, $r_{12} = r_{13} = r_{23} = a$, $r_{14} = r_{24} = r_{34} = (1/\sqrt{3})\, a$.

$$U = k\left(\dfrac{q_1 q_2}{r_{12}} + \dfrac{q_1 q_3}{r_{13}} + \dfrac{q_2 q_3}{r_{23}}\right) + k\left(\dfrac{q_1 q_4}{r_{14}} + \dfrac{q_2 q_4}{r_{24}} + \dfrac{q_3 q_4}{r_{34}}\right)$$

$$U = k\left(3\dfrac{q^2}{a}\right) + k\left(3\dfrac{qQ}{\left(\dfrac{1}{\sqrt{3}}\right)a}\right)$$

$$U = 3k(1 + \sqrt{3})\dfrac{q}{a}(q + Q)$$

$$U = \dfrac{3(1 + \sqrt{3})\, q\, (q + Q)}{4\pi\varepsilon_0\, a}$$

94. (a)

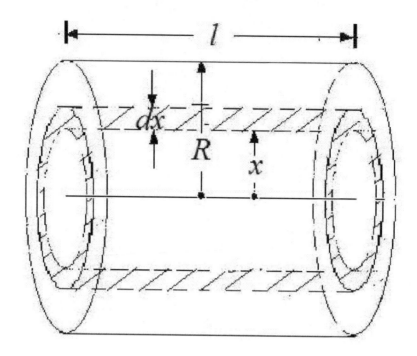

Consider a co-axial cylindrical shell of radius x and thickness dx and of length l. The volume of this shell is
$$dV = 2\pi x\, dx\, l$$
The charge on the shell
$$dq = 2\pi x\, dx\, l\, \rho(x)$$
$$= 2\pi x\, dx\, l\, \rho_0 x^2$$
Since charge density is zero outside the cylinder, the charge contained in length l of the cylinder for $r > R$ is
$$q = 2\pi l \rho_0 \int_0^R x^3 dx$$
$$= 2\pi l \rho_0 \frac{R^4}{4}$$

(b)

First note that symmetry considerations tell us that, for a long cylinder, the electric field is everywhere radial in the plane cutting the cylinder normally. We make use of Gauss's law to calculate electric field at a radial distance $r\ (< R)$.

For that, we take cylindrical Gaussian surface S of radius r and length l. The radial electric field being parallel to the end caps of surface S, the electric flux through each end cap is zero. On the cylindrical part of S, E is normal to the surface at every point and its magnitude is constant. The surface area of the curved part is $2\pi rl$.

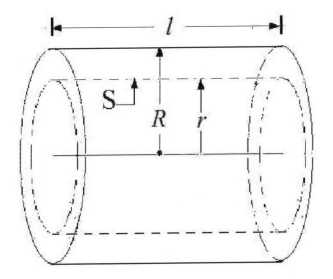

Flux through S is

φ = Flux through the cylindrical surface of S = $E \times 2\pi r\, l$

To find the charge enclosed in S, we make use of arguments similar to those in part (a). We have

$$q_{enclosed\ in\ S} = 2\pi\, l\, \rho_0 \int_0^r x^3 dx = 2\pi\, l\, \rho_0 \frac{r^4}{4}$$

By Gauss's law,

$$\varphi = \frac{q_{enclosed\ in\ S}}{\varepsilon_0}$$

$$E \times 2\pi r l = \frac{2\pi\, l\, \rho_0 r^4}{4\, \varepsilon_0}$$

$$E = \frac{2\pi\, l\, \rho_0 r^4}{8\pi r\, l\, \varepsilon_0}$$

$$E(r) = \frac{\rho_0 r^3}{4\, \varepsilon_0},\ (r < R)$$

(c)

To find the electric field **E** at a radial distance $r \, (> R)$, we consider a cylindrical Gaussian surface S' of radius r and length l.

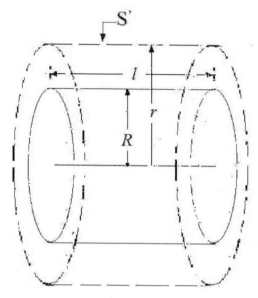

The field being radial, the electric flux through each end cap is zero. As in part (b) above, flux through S' is

$\varphi = E \times 2\pi r \, l$

In this case, the charge enclosed in S' equals the charge q contained in length l of the cylinder. We have already calculated its value in part (a) above. Therefore the use of Gauss's law

$$\varphi = \frac{q_{enclosed \, in \, S'}}{\varepsilon_0} = \frac{q}{\varepsilon_0}$$

$$E(r) \times 2\pi r l = \frac{2\pi \, l \, \rho_0 R^4}{4 \, \varepsilon_0}$$

$$E(r) = \frac{2\pi \, l \, \rho_0 R^4}{8\pi r \, l \, \varepsilon_0}$$

$$E(r) = \frac{\rho_0 R^4}{4 \, \varepsilon_0 r} \, , (r > R)$$

(d)

Substitute $r = R$ in expressions for electric field obtained in part (b), we obtain

$$E(r = R) = \frac{\rho_0 R^3}{4 \, \varepsilon_0} \quad \ldots (i)$$

Putting $r = R$ in the expressions for electric field obtained in part (c), we get

$$E(r = R) = \frac{\rho_0 R^4}{4 \, \varepsilon_0 R} = \frac{\rho_0 R^3}{4 \, \varepsilon_0} \quad \ldots (ii)$$

We find that the value of electric field at $r = R$ (i.e., on the surface of the cylinder) coincide. This is because the charge density is not infinite anywhere and hence the electric field must be continuous.

95. (a)

In the given graph, volume charge density $\rho(r)$ is plotted along y-axis and radial distance r is plotted along x-axis. The intercept on y-axis is ρ_0 and the slope is $(-\rho_0/R)$. So, the equation of the line is

$$\rho(r) = -\frac{\rho_0}{R}r + \rho_0$$

To find the electric field **E** at a radial distance r ($> R$), note that **E** is radial since its magnitude depends only on the radial distance r from the center O of the sphere.

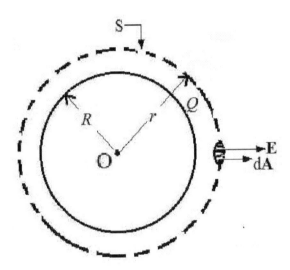

We consider a spherical Gaussian surface S of radius r centered at O. **E** and the area element d**A** at every point of S are parallel and the flux through each element is $E\,dA \cos 0° = E\,dA$. Summing over all area elements, the flux through the Gaussian surface is

$$\varphi = \oint \mathbf{E} \cdot \mathbf{dA} = E \times (4\pi r^2)$$

Since the charge density is zero outside the sphere, the charge enclosed in S equals the total charge Q in the sphere.

Use of Gauss's law

$$\varphi = \frac{q_{enclosed}}{\varepsilon_0}$$

gives

$$E \times (4\pi r^2) = \frac{Q}{\varepsilon_0}$$

$$E(r) = \frac{Q}{4\pi \varepsilon_0 r^2}, r > R$$

(b)

To find electric field at a radial distance $r\ (< R)$, we again consider a Gaussian spherical surface S' with center O and radius r.

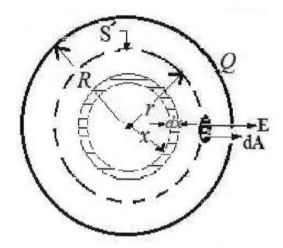

The field being radial, the electric flux through S' (as in part (a) above) is

$$\varphi = \oint \mathbf{E}\cdot \mathbf{dA} = E \times (4\pi r^2)$$

To calculate the charge enclosed in S', we consider a spherical shell of radius x, thickness dx and centered at O. The volume of this shell is

$$dV = 4\pi x^2\, dx$$

The charge on the shell

$$dq = 4\pi x^2\, dx\, \rho(x) = 4\pi x^2\, dx \left(-\frac{\rho_0}{R}x + \rho_0\right)$$

So, the charge enclosed in S' is

$$q = 4\pi \int_0^r \left(-\frac{\rho_0}{R}x + \rho_0\right) x^2\, dx$$

$$q = 4\pi \left(-\frac{\rho_0}{R}\frac{x^4}{4}\bigg|_{x=0}^{x=r} + \rho_0 \frac{x^3}{3}\bigg|_{x=0}^{x=r}\right)$$

$$q = 4\pi \rho_0 r^3 \left(-\frac{r}{4R} + \frac{1}{3}\right) = \frac{\pi \rho_0 r^3}{3R}(4R - 3r)$$

By Gauss's law,

$$\varphi = \frac{q_{enclosed}}{\varepsilon_0}$$

$$E \times (4\pi r^2) = \frac{q}{\varepsilon_0} = \frac{\pi \rho_0 r^3}{3\varepsilon_0 R}(4R - 3r)$$

$$E = \frac{\rho_0 r}{12\varepsilon_0 R}(4R - 3r),\ (r < R)$$

(c)

As in part (b), the charge on a spherical shell of radius x, thickness dx and centered at O is

$$dq = 4\pi x^2 \, dx \, \rho(x) = 4\pi x^2 \, dx \left(-\frac{\rho_0}{R}x + \rho_0\right)$$

By integrating dq from $x=0$ to $x=R$, we should obtain the total charge in the sphere of radius R. So,

$$Q = 4\pi \int_0^R \left(-\frac{\rho_0}{R}x + \rho_0\right) x^2 \, dx$$

$$Q = 4\pi \left(-\frac{\rho_0}{R} \frac{x^4}{4}\bigg|_{x=0}^{x=R} + \rho_0 \frac{x^3}{3}\bigg|_{x=0}^{x=R}\right)$$

$$Q = \frac{\pi \rho_0 R^3}{3}$$

$$\rho_0 = \frac{3Q}{\pi R^3}$$

(d)

We can calculate the electric potential at the center O of the sphere by using the formula

$$V_O = -\int_{r=\infty}^{r=0} E \, dr$$

We break the above integrals into two integrals and make use of the electric field values in parts (a) and (b) as follows.

$$\int_{r=\infty}^{r=0} E \, dr = \int_{r=\infty}^{r=R} E \, dr + \int_{r=R}^{r=0} E \, dr$$

Using the value of E in part (a),

$$\int_{r=\infty}^{r=R} E \, dr = \int_{r=\infty}^{r=R} \frac{Q}{4\pi \varepsilon_0 r^2} \, dr = \frac{Q}{4\pi \varepsilon_0} \int_{r=\infty}^{r=R} \frac{1}{r^2} \, dr = \frac{Q}{4\pi \varepsilon_0} |-r^{-1}|_{r=\infty}^{r=R}$$

$$\int_{r=\infty}^{r=R} E \, dr = -\frac{Q}{4\pi \varepsilon_0 R}$$

Also,

$$\int_{r=R}^{r=0} E \, dr = \frac{\rho_0}{12 \varepsilon_0 R} \int_{r=R}^{r=0} (4Rr - 3r^2) \, dr$$

Substituting the value of ρ_0 in terms of Q, we have

$$\int_{r=R}^{r=0} E \, dr = \frac{Q}{4\pi \varepsilon_0 R^4} \int_{r=R}^{r=0} (4Rr - 3r^2) \, dr = \frac{Q}{4\pi \varepsilon_0 R^4} \left[2R \, |r^2|_{r=R}^{r=0} - |r^3|_{r=R}^{r=0}\right]$$

$$\int_{r=R}^{r=0} E\,dr = -\frac{Q}{4\pi\varepsilon_0 R^4}[2R^3 - R^3] = -\frac{Q}{4\pi\varepsilon_0 R}$$

Substituting these values in the equation for the electric potential at the center O, we get

$$V_O = -\int_{r=\infty}^{r=0} E\,dr$$

$$V_O = -\int_{r=\infty}^{r=R} E\,dr - \int_{r=R}^{r=0} E\,dr$$

$$V_O = \frac{Q}{4\pi\varepsilon_0 R} + \frac{Q}{4\pi\varepsilon_0 R} = \frac{Q}{2\pi\varepsilon_0 R}$$

(e)

We know that the charge on a spherical shell of radius x, thickness dx and centered at O is

$$dq = 4\pi x^2\,dx\,\rho(x) = 4\pi x^2\,dx\left(-\frac{\rho_0}{R}x + \rho_0\right)$$

Every point on this shell is at the same distance x from O. So, the potential due to this shell is

$$dV = k\frac{dq}{x} = k\frac{4\pi x^2\,dx\left(-\frac{\rho_0}{R}x + \rho_0\right)}{x} = \frac{1}{\varepsilon_0}\left(-\frac{\rho_0}{R}x^2 + \rho_0 x\right)$$

Since the charge density is zero outside the sphere, the electric potential at the center O is obtained by integrating the above from $x = 0$ to $x = R$.

$$V_O = \frac{1}{\varepsilon_0}\left(-\frac{\rho_0}{R}\int_{x=0}^{x=R} x^2\,dx + \rho_0\int_{x=0}^{x=R} x\,dx\right)$$

$$V_O = \frac{1}{\varepsilon_0}\left(-\frac{\rho_0}{R}\frac{R^3}{3} + \rho_0\frac{R^2}{2}\right) = \frac{\rho_0 R^2}{6\varepsilon_0}$$

Substituting the value of ρ_0 in terms of Q,

$$V_O = \frac{\left(\frac{3Q}{\pi R^3}\right)R^2}{6\varepsilon_0} = \frac{Q}{2\pi\varepsilon_0 R}$$

Thus, we get the same result for the electric potential at O as that in part (d).

96. (a)

We shall divide the disc into concentric rings and then add up the contributions of all the rings.

As shown in the figure, we consider a ring of radius r and small radial width dr centered at O.

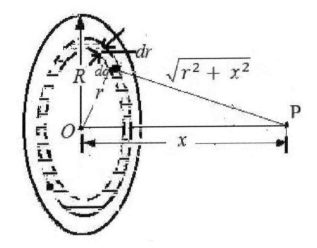

Area of the ring = $2\pi r dr$

Charge on the ring, $dq = (2\pi r dr)\sigma$

The electric potential at P due to dq is

$$dV = k\frac{dq}{\sqrt{r^2+x^2}}$$

$$dV = k\frac{(2\pi r dr)\sigma}{\sqrt{r^2+x^2}}$$

Since every point on this ring is at the same distance $(\sqrt{r^2+x^2})$ from P, we can find the potential V at P due to the charge on the whole disc by integrating dV from $r = 0$ to $r = R$. So,

$$V = k(2\pi\sigma)\int_{r=0}^{r=R}\frac{rdr}{(r^2+x^2)^{\frac{1}{2}}}$$

Substituting $k = \dfrac{1}{4\pi\varepsilon_0}$,

$$V = \frac{\sigma}{2\varepsilon_0}\int_{r=0}^{r=R}\frac{rdr}{(r^2+x^2)^{1/2}}$$

Put

$$z = (r^2+x^2)$$

Differentiating with respect to r

$$dz = 2r\,dr$$

$$V = \frac{\sigma}{4\varepsilon_0}\int_{z=x^2}^{z=(R^2+x^2)}\frac{dz}{z^{\frac{1}{2}}} = \frac{\sigma}{2\varepsilon_0}\left(z^{\frac{1}{2}}\Big|_{z=x^2}^{z=(R^2+x^2)}\right)$$

$$V = \frac{\sigma}{2\varepsilon_0}\left((R^2+x^2)^{\frac{1}{2}} - x\right)$$

(b)

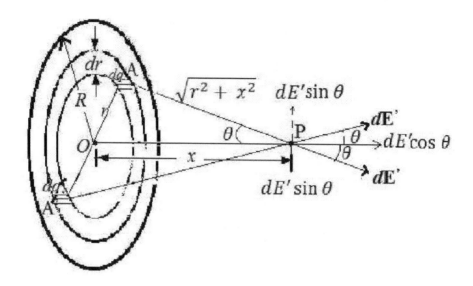

As in part (b), we again consider a ring of radius r and small radial width dr centered at O. A small element of charge dq' on the ring (shown shaded) at A produces electric field $d\mathbf{E}'$ at P that can be decomposed into two components: one perpendicular to and the other along OP. The component perpendicular to OP is cancelled by another perpendicular component of same magnitude at a diametrically opposite element at A' as shown but the components along OP add up. Such consideration holds for each small element on the ring. We need to add only the component along OP. [The above result is consistent with symmetry considerations which tell us that the only unique direction is perpendicular to the plane of the disk along OP and so the net electric field should lie along OP.]

Now, as shown, if θ is the angle APO, the component along OP due to dq' is

$$dE' = k\frac{dq'}{(r^2 + x^2)}\cos\theta = k\frac{dq'}{(r^2 + x^2)}\frac{x}{\sqrt{r^2 + x^2}}$$

Since every element of charge on the ring of radius r is at the same distance $\left(\sqrt{r^2 + x^2}\right)$ from P, the electric field at P due to whole ring is

$$dE = k\frac{x}{(r^2 + x^2)^{\frac{3}{2}}}\oint dq'$$

$$dE = k\frac{x}{(r^2 + x^2)^{\frac{3}{2}}}(2\pi r dr\sigma)$$

, where $(2\pi r dr\sigma)$ is the total charge on the ring.

Having found the contribution of the chosen ring, we can find the magnitude of electric field due to whole disc by integrating the above obtained result from $r = 0$ to $r = R$.

$$E = k\,(2\pi\sigma) \int_{r=0}^{r=R} \frac{r\,dr}{(r^2 + x^2)^{\frac{3}{2}}}$$

To carry out the integration, we again substitute

$$z = (r^2 + x^2)$$

Differentiating with respect to r

$$dz = 2\,r\,dr$$

$$E = k\,(\pi\sigma) \int_{z=x^2}^{z=(R^2+x^2)} \frac{dz}{z^{\frac{3}{2}}} = k\,(\pi\sigma) \left. \frac{z^{-\frac{1}{2}}}{-\frac{1}{2}} \right|_{z=x^2}^{z=(R^2+x^2)}$$

$$E = k\,(2\pi\sigma) \left(\frac{1}{x} - \frac{1}{(R^2 + x^2)^{\frac{1}{2}}} \right)$$

$$E = \frac{\sigma}{2\,\varepsilon_0} \left(1 - \frac{x}{(R^2 + x^2)^{\frac{1}{2}}} \right)$$

(c)

We begin with the expression of electric potential at P obtained in part (a).

$$V = \frac{\sigma}{2\,\varepsilon_0} \left((R^2 + x^2)^{\frac{1}{2}} - x \right)$$

We differentiate the above expression with respect to x

$$\frac{dV}{dx} = \frac{\sigma}{2\,\varepsilon_0} \left(\frac{d(R^2 + x^2)^{\frac{1}{2}}}{dx} - 1 \right) = \frac{\sigma}{2\,\varepsilon_0} \left\{ \frac{1}{2} \frac{2x}{(R^2 + x^2)^{\frac{1}{2}}} - 1 \right\}$$

$$\frac{dV}{dx} = \frac{\sigma}{2\,\varepsilon_0} \left(\frac{x}{(R^2 + x^2)^{1/2}} - 1 \right)$$

We thus find that $\left(-\dfrac{dV}{dx}\right)$ agrees with the expression of electric field E obtained in part (b). In other words, the electric field in the direction of increasing x is

$$E_x = -\frac{dV}{dx}$$

Recall that the electric field calculated in part (b) is along OP in the direction of increasing x.

(d)

(i) When $x \gg R$, the distance of point P where electric field is to be calculated, is much larger than the size of the disk. Then the charged disk can be approximated as a point charge. We have

$$E = \frac{\sigma}{2\varepsilon_0}\left(1 - \frac{x}{(R^2+x^2)^{\frac{1}{2}}}\right) = \frac{\sigma}{2\varepsilon_0}\left(1 - x(R^2+x^2)^{-\frac{1}{2}}\right)$$

$$E = \frac{\sigma}{2\varepsilon_0}\left\{1 - \left(1 + \frac{R^2}{x^2}\right)^{-\frac{1}{2}}\right\}$$

Note that $x \gg R$ implies that $\frac{R^2}{x^2} \ll 1$. Then we can use the binomial approximation

$(1+y)^\alpha \approx 1 + \alpha y$

$$E \approx \frac{\sigma}{2\varepsilon_0}\left\{1 - \left(1 - \left(\frac{1}{2}\right)\frac{R^2}{x^2}\right)\right\}$$

$$E = \frac{\sigma(4\pi R^2)}{4\pi\varepsilon_0 x^2}$$

Note that $(4\pi R^2 \sigma)$ is the total charge (say) Q on the disk.

$$E = \frac{Q}{4\pi\varepsilon_0 x^2}$$

The above result agrees with the magnitude of the electric field due to a point charge Q at O at a distance x from the charge. Its direction is radially outward along OP.

(ii) When $R \gg x$, the size of the disk is much larger than the distance of point P where electric field is to be calculated. In this case, we have

$R^2 + x^2 \approx R^2$

and so

$$E = \frac{\sigma}{2\varepsilon_0}\left(1 - \frac{x}{(R^2+x^2)^{1/2}}\right) \approx \frac{\sigma}{2\varepsilon_0}\left(1 - \frac{x}{R}\right)$$

Since $\frac{x}{R} \ll 1$

$$E = \frac{\sigma}{2\varepsilon_0}$$

This result coincides with the magnitude of electric field near a very large non-conducting sheet having a uniform surface charge density σ. The direction of electric field is perpendicular to the sheet along OP.

97. (a)

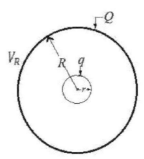

To find the electric potential V_R at the surface of the spherical conducting shell, recall that for points outside and on the shell, the electric potential is as if the entire charge of the shell is concentrated at its center of the shell. The surface of the conductor being an equipotential surface, the charge Q given to the spherical metallic shell spreads uniformly on its surface. Inside the shell, the electric potential is constant and has the same value as on the surface of the shell. As for the small conducting sphere, the electric potential at points outside and on the sphere also equal to that of the entire charge of the sphere at the center of the sphere.

In view of the above results, the electric potential at the surface of the spherical conducting shell due to charge Q on the shell is $\left(k\dfrac{Q}{R}\right)$ and that due to charge q on the sphere is $\left(k\dfrac{q}{R}\right)$. So, the net potential at the surface of the shell has the value

$$V_R = k\frac{Q}{R} + k\frac{q}{R} = \frac{1}{4\pi\varepsilon_0}\left(\frac{Q}{R}+\frac{q}{R}\right)$$

(b)

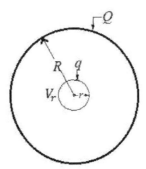

The electric potential at the surface of the sphere due to charge q on the sphere

$$= \left(k\frac{q}{r}\right)$$

Noting that the sphere is inside the shell, the potential at the surface of the sphere due to charge Q on the shell

$$= k\frac{Q}{R}$$

Considering the contributions of both the charges, we have for the total potential at the surface of the sphere

$$V_r = k\frac{q}{r} + k\frac{Q}{R} = \frac{1}{4\pi\varepsilon_0}\left(\frac{q}{r}+\frac{Q}{R}\right)$$

(c)

Let us calculate the potential difference between the surface of the sphere and that of the shell. Using the values in parts (a) and (b), we have

$$V_r - V_R = \frac{1}{4\pi\varepsilon_0}\left(\frac{q}{r} + \frac{Q}{R} - \frac{q}{R} - \frac{Q}{R}\right) = \frac{q}{4\pi\varepsilon_0}\left(\frac{R-r}{rQ}\right)$$

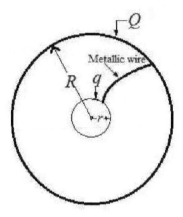

We find that $(V_r - V_R)$ is independent of the charge Q on the outer shell. Moreover, given that q is positive and $R > r$, we find that $(V_r - V_R)$ is positive. In other words, the inner sphere is at a higher potential than the outer shell. Remember that positive charge flows from higher to lower potential. Hence, on connecting the inner sphere to the outer shell with a metallic wire, irrespective of charge on the shell (be it positive or negative), any charge given to the sphere will flow to the outer shell.

(d)

This is because more charge leads to increase of the electric field E ($E = k\frac{Q}{R^2}$, or $E \propto Q$) near the shell. Intense electric field can cause detachment of outer electrons from the parent atoms of the air surrounding the shell and the air starts behaving like a conductor. Such a phenomenon is known as dielectric breakdown of the air. When it happens, the charge on the shell begins to leak to air and no more charge can be given to it.

(e)

For charge Q on the surface of the spherical shell of radius R electric field at its surface is

$$E = \frac{Q}{4\pi\varepsilon_0 R^2}$$

Electric potential at its surface is

$$V = \frac{Q}{4\pi\varepsilon_0 R}$$

Division gives

$$\frac{V}{E} = R$$

So,

$V = E\,R$

Substituting the values of E and R, we get

$V = \left(3 \times 10^8 \; \frac{V}{m}\right)(0.5\; m) = 1.5 \times 10^8$ V.

98. (a)

In the given network, C_1 and C_2 are connected in series.

The effective capacitance C'_s of these two capacitors is given by

$$\frac{1}{C'_s} = \frac{1}{C_1} + \frac{1}{C_2} = \frac{C_1 + C_2}{C_1 C_2}$$

$$C'_s = \frac{C_1 C_2}{C_1 + C_2} = \frac{(3\; \mu F)(6\; \mu F)}{(3\; \mu F + 6\; \mu F)} = 2\; \mu F$$

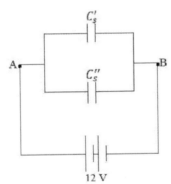

C_3 and C_4 are also connected in series. Their effective capacitance C''_s is

$$C''_s = \frac{C_3 C_4}{C_3 + C_4} = \frac{(2\; \mu F)(4\; \mu F)}{(2\; \mu F + 4\; \mu F)} = \left(\frac{4}{3}\right) \mu F$$

Now, C'_s and C''_s are connected in parallel. So, the equivalent capacitance between A and B is

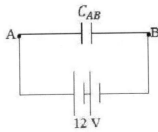

$$C_{AB} = C'_s + C''_s = 2\; \mu F + \left(\frac{4}{3}\right) \mu F = \left(\frac{10}{3}\right) \mu F$$

(b)

C_{AB} is joined across a battery of voltage $V = 12$ V. Therefore, total charge supplied by the battery is

$$Q_T = C_{AB} \times V = \left(\frac{10}{3}\right) \mu F \times 12\, V = 40\, \mu Q$$

(c)

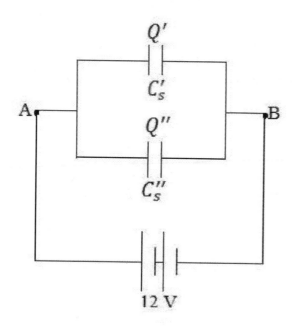

Let Q' and Q'' be the charges on the series combinations C'_s and C''_s. Since C'_s and C''_s are in parallel, they have a common potential,

$$\frac{Q'}{C'_s} = \frac{Q''}{C''_s} = \frac{Q' + Q''}{C'_s + C''_s} \quad \ldots\ldots\ldots (i)$$

Conservation of charge gives

$$Q_T = Q' + Q'' \quad \ldots\ldots\ldots (ii)$$

From Eqs. (i) and (ii),

$$Q' = \frac{Q_T}{C_{AB}} C'_s = \frac{40\, \mu Q}{\left(\frac{10}{3}\right) \mu F} \cdot 2\, \mu F = 24\, \mu Q$$

and $Q'' = \dfrac{Q_T}{C_{AB}} C''_s = \dfrac{40\, \mu Q}{\left(\frac{10}{3}\right) \mu F} \left(\dfrac{4}{3}\right) \mu F = 16\, \mu Q$

Let Q_1, Q_2, Q_3 and Q_4 be the charges on C_1, C_2, C_3 and C_4 respectively. Since C_1 and C_2 are connected in series, charge across each of them is Q'. Similarly, charge across each of C_3 and C_4 is Q''. So,

$$Q_1 = Q_2 = 24\, \mu Q$$

$Q_3 = Q_4 = 16 \mu Q$

(d)

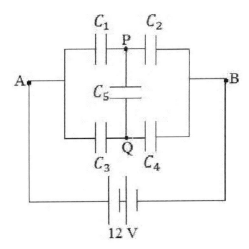

(i) To find the charge on C_5, we need to find the potential difference $(V_P - V_Q)$ between points P and Q.

We have potential difference V_1 between the plates of C_1

$$V_1 = V_A - V_P = \frac{Q_1}{C_1} = \frac{24 \mu Q}{3 \mu F} = 8 V$$

Potential difference V_1 between the plates of C_1

$$V_2 = V_P - V_B = \frac{Q_2}{C_2} = \frac{24 \mu Q}{6 \mu F} = 4 V$$

Similarly,

$$V_3 = V_A - V_Q = \frac{Q_3}{C_3} = \frac{16 \mu Q}{2 \mu F} = 8 V$$

$$V_4 = V_Q - V_B = \frac{Q_3}{C_3} = \frac{16 \mu Q}{4 \mu F} = 4 V$$

$V_1 - V_2 = V_A - 2 V_P + V_B = 4 V$ ....(i)

$V_3 - V_4 = V_A - 2 V_Q + V_B = 4 V$ ....(ii)

Subtracting Eq. (i) from Eq. (ii),

$2 (V_P - V_Q) = 0$

$V_P - V_Q = 0$

Since the potential difference between points P and Q is zero, the charge on C_5 joined between these two points will be zero.

(ii) There will be no change in the values of various quantities in parts (a), (b) and (c). This is because no charge flows between P and Q and hence the potential and charge values at all other points remain unchanged.

99. (a)

Note that by spherical symmetry, the electric **E** at any point between the spheres is radially outward and has the same magnitude E at all points that are at the same distance r from the center.

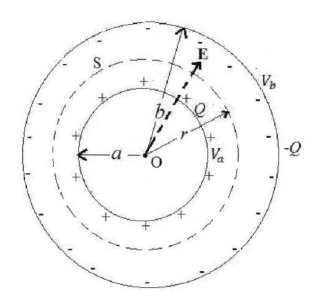

We take the Gaussian spherical surface S with the center O of the concentric shells and having radius r. The flux φ through the Gaussian surface is

$\varphi = \oint \mathbf{E} \cdot \mathbf{dS} = E \oint dS = E \times (4\pi r^2)$

Since the charge Q is contained inside S, application of Gauss's law gives

$\varphi = \dfrac{q_{enclosed}}{\varepsilon_0}$

$E \times (4\pi r^2) = \dfrac{Q}{\varepsilon_0}$

$E = \dfrac{Q}{4\pi\varepsilon_0 r^2}$

The potential difference between the two conductors is

$V = V_a - V_b = -\displaystyle\int_b^a E\, dr$

Substituting the value of E,

$V = -\dfrac{Q}{4\pi\varepsilon_0}\displaystyle\int_b^a \dfrac{dr}{r^2} = -\dfrac{Q}{4\pi\varepsilon_0}\left(\dfrac{r^{-1}}{-1}\Big|_{r=b}^{r=a}\right)$

$V = -\dfrac{Q}{4\pi\varepsilon_0}\left(\dfrac{1}{b} - \dfrac{1}{a}\right) = \dfrac{Q(b-a)}{4\pi\varepsilon_0\, ab}$

(b)

We can use the result above to find the capacitance of the spherical capacitor. By definition, capacitance C is

$$C = \frac{Q}{V} = \frac{4\pi\varepsilon_0 \, ab}{(b-a)}$$

(c)

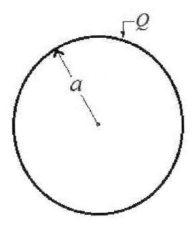

We know that charge Q placed on a conducting sphere of radius a spreads uniformly on its surface. The electric potential V at points outside and on its surface is equal to that of the point charge Q at the center of the sphere.

$$V = \frac{Q}{4\pi\varepsilon_0 \, a}$$

Therefore, the capacitance of an isolated spherical conductor is

$$C = \frac{Q}{V} = 4\pi\varepsilon_0 \, a$$

(d)

(i) When the radius b of the outer shell becomes infinity, we can say that we are dealing with an isolated conductor. We have

$$C = \frac{4\pi\varepsilon_0 \, ab}{(b-a)}$$

Under this condition of $b \to \infty$, we have

$$C \approx \frac{4\pi\varepsilon_0 \, ab}{b} = 4\pi\varepsilon_0 \, a$$

The above result agrees with the result for the capacitance of a single spherical conductor.

(ii) When both a and b become very large and $(b - a) = d$ (say) is kept fixed, we have the case of a parallel-plate capacitor.

Under these conditions

$$C = \frac{4\pi\varepsilon_0\, ab}{(b-a)} \approx \frac{4\pi\varepsilon_0\, ab}{d}$$

Again, $4\pi\, ab \approx 4\pi a^2 = A$

where a is the radius of each plate and A is the area. So C then becomes

$$C = \frac{\varepsilon_0 A}{d}$$

The above equation agrees with the result for the capacitance of a parallel plate capacitor.

(e)

The equation for the capacitance of an isolated spherical conductor ($C = 4\pi\varepsilon_0\, a$) gives the result for the radius a

$$a = \frac{C}{4\pi\varepsilon_0}$$

Substitute $C = 1$ F and

$\left(\frac{1}{4\pi\varepsilon_0}\right) = 9 \times 10^9$ N^{-1} m^{-1}C^2.

Noting that in SI units, $1\,F = 1\,CV^{-1} = 1\,C\,(N\,mC^{-1})^{-1} = 1\,N^{-1}\,m^{-1}C^2$, we have

$a = 9 \times 10^9 (N^{-1}\,m^{-1}C^2)(Nm^2C^{-2})$

$a = 9 \times 10^9\ m$

We thus find that the radius of an isolated spherical conductor that has a capacitance of 1 F is more than 1000 times the radius of earth.

The above result leads us to conclude that, 1 F is a very large unit. No wonder the commonly used units in practice are its sub-multiples like $1\,\mu F = 10^{-6}$ F, $1\,n F = 10^{-9}$ F and $1\,pF = 10^{-12}$ F.

100. (a)

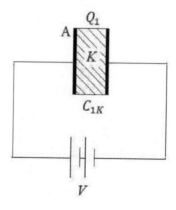

When a dielectric slab having dielectric constants K is introduced to fill the space between its plates, the capacitance of the capacitor A becomes

$C_{1K} = K C_0$

When capacitor A with dielectric slab is charged to potential V, the charge on the capacitor is

$Q_1 = C_{1K} V$

Electrostatic energy stored in A with dielectric slab is

$U_{1K} = \dfrac{1}{2} \dfrac{Q_1^2}{C_{1K}} = \dfrac{1}{2} \dfrac{Q_1^2}{K C_0}$

(b)

When battery is disconnected, the charge on the capacitor A $(= Q_1)$ continues to be remain unchanged. Also, when the dielectric slab is removed, the capacitance of A is C_0.

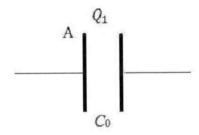

The changed value of stored energy becomes

$U_1' = \dfrac{1}{2} \dfrac{Q_1^2}{C_0}$

Using the result obtained in part (a), let us calculate the change in the stored energy.

$U_1' - U_{1K} = \dfrac{1}{2} \dfrac{Q_1^2}{C_0} - \dfrac{1}{2} \dfrac{Q_1^2}{K C_0}$

$= \dfrac{1}{2} \dfrac{Q_1^2}{C_0} \left(1 - \dfrac{1}{K}\right)$

$U_1' - U_{1K} = \dfrac{1}{2} \dfrac{Q_1^2}{C_0 K} (K - 1)$

Since $K > 1$, we find that when dielectric slab is removed, the stored electrostatic energy increases. This is because the electric field between the plates attracts the slab of dielectric into the capacitor. An external agent has to do work in removing the slab. By law of conservation of energy, such wok done is converted into the stored energy of the capacitor.

(c)

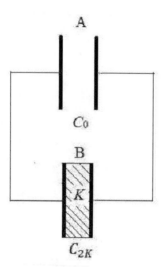

When the same slab of dielectric is inserted between the plates of capacitor B, its capacitance becomes

$C_{2K} = K\, C_0$

The capacitance of capacitor A without the dielectric is C_0

In the given figure, two capacitors A (without the slab) and B (with the slab) are seen to be joined in parallel. So, the equivalent capacitance of the parallel combination is

$C_p = C_0 + C_{2K} = C_0 + K\, C_0 = C_0(1 + K)$

$C_p = C_0(1 + K)$

(d)

Since the battery remains disconnected, the charge on the equivalent capacitance C_p continues to be Q_1.

Therefore, the electrostatic energy stored in the combination of two capacitors is

$U_p = \dfrac{1}{2}\dfrac{Q_1^{\,2}}{C_p} = \dfrac{Q_1^{\,2}}{2\, C_0(1 + K)}$

Made in the USA
Columbia, SC
02 November 2023